VCSELs for Cesium-Based Miniaturized
Atomic Clocks

DISSERTATION

to obtain the academic degree of

DOKTOR-INGENIEUR
(DR.-ING.)

from the Faculty of Engineering and Computer Science
Ulm University

by

Ahmed Al-Samaneh
from Amman, Jordan

Referees: apl. Prof. Dr.-Ing. Rainer Michalzik
  Prof. Dr. Martin Hofmann
Dean of the Faculty: Prof. Dr. Tina Seufert

Ulm, March 20th, 2014

Herstellung und Verlag:
BoD – books on demand, Norderstedt.
ISBN 978-3-7386-1366-7

*"Whoever treads a path seeking knowledge,
Allah will make easy for him the path to Paradise",
The Prophet Muhammad (PBUH).*

Acknowledgement

This dissertation would not have been possible without the help of many people. Firstly, I would like to express my sincere gratitude to my supervisor apl. Prof. Dr.-Ing. Rainer Michalzik, who gave me this great opportunity to do this work and to participate in a European collaborative research project involving several academic partners, international research institutes and many industrial enterprises with the target to realize the first European miniaturized atomic clock demonstrators. He was always a source of support and new ideas that have had a profound effect on this dissertation.

Special thanks go to all people at the Institute of Optoelectronics which I have gladly worked with during this dissertation, particularly, Wolfgang Schwarz who taught me how to process VCSELs, and always supported me with his vast expertise in microwave measurements, Dietmar Wahl who put much effort in growing many VCSEL wafers where reaching the Cs D_1 wavelength was always challenging, my friend and office-mate Alexander Kern for his valuable advices to improve measurement setups, helping with SEM pictures and of course the very enjoyable time we spent together during our PhDs, Rudolf Rösch for his continuous support in dry etching, contact metallization and substrate thinning, Susanne Menzel for her great help in dealing with different chemicals in the cleanroom, Dr.-Ing. Jürgen Mähnß for reading and correcting the first draft of this dissertation, Alexander Hein for assistance with lifetime measurements and Anna Bergmann for her kind help with contact metallization.

I also thank very much all students who worked with me during the years of this PhD work and contributed strongly with their diploma, master and bachelor theses to the success of the atomic clock VCSELs. Namely, my thanks go to Simeon Renz, Andreas Strodl, Md. Jarez Miah and Mustafa Kazu. I would also like to thank Md. Tanvir Haidar, Sujoy Paul and Niazul Islam Khan for their great help in characterizing atomic clock VCSELs through their student jobs at the Institute of Optoelectronics.

I am very grateful to all people at the Institute of Electron Devices and Circuits at Ulm University for their support and assistance, particularly, Yakiv Men for performing the electron-beam lithography steps for all the grating VCSEL wafers and Tatyana Purtova and Gang Liu for their continuous assistance with microwave reflection measurements.

Many thanks go to all the project partners of MAC-TFC, especially Dr. Christoph Affolderbach from UniNE-LTF and Dr. Vincent Giordano from FEMTO-ST, who answered all the questions on the physics of miniaturized atomic clocks, enhancing greatly my understanding of the subject and Dr. Rahel Strässle and Dr. Yves Pétremand from SAMLAB at EPFL for providing me with microfabricated Cs vapor cells for the measurement of the absorption spectra presented in App. G.

I am very grateful to Dr.-Ing. Dieter Wiedenmann from Philips U-L-M Photonics and his team for their support especially with mounting VCSEL chips in TO cans. Acknowledgements also go to Dr. Pierluigi Debernardi from IEIIT-CNR Torino for performing simulations of grating VCSELs using his excellent model, Dr.-Ing. Marwan Bou Sanayeh from Notre Dame University who helped very much by processing VCSELs during his frequent visits as a guest scientist at the Institute of Optoelectronics.

My gratitude is also extended to the members of the PhD commission, apl. Prof. Dr.-Ing. Rainer Michalzik, Prof. Dr. Martin Hofmann from Ruhr-Universität Bochum, Prof. Dr. Ulrich Herr and Prof. Dr.-Ing. Albrecht Rothermel for all the efforts they exerted.

Finally, special thanks and appreciation to my beloved parents, my dear brother Samer and my dear sister Soha for their unconditional support and encouragement despite the long distance.

This dissertation was first published by Ulm University, Ulm in 2014.

Contents

1 Introduction and Motivation   **1**

2 Clocks and Frequency Standards: Concept, History, Miniaturization, and Applications   **5**
- 2.1 Key Aspects of Clocks 5
  - 2.1.1 Accuracy 6
  - 2.1.2 Stability 6
  - 2.1.3 Quality Factor 7
- 2.2 Historical Prospective of Clocks 7
  - 2.2.1 Mechanical Clocks 8
  - 2.2.2 Quartz-Based Clocks 9
  - 2.2.3 Microwave Atomic Clocks 10
- 2.3 Miniaturization of Atomic Clocks 11
  - 2.3.1 Three-Level System and CPT Spectroscopy 13
  - 2.3.2 Microfabricated Alkali Vapor Cells 15
  - 2.3.3 Frequency Stability of MACs 16
  - 2.3.4 Requirements on the Laser Source 18
- 2.4 Applications of MACs 19

3 Fundamentals of VCSELs   **21**
- 3.1 Device Structure and Properties 21
- 3.2 Threshold Conditions 25
- 3.3 Operation Characteristics 27
- 3.4 Temperature Behavior 29
  - 3.4.1 Red-Shift Effect 29
  - 3.4.2 Thermal Resistance 30
- 3.5 Dynamic and Noise Behavior 33

    3.5.1   Rate Equations 33
    3.5.2   Small-Signal Modulation Response 34
    3.5.3   Intensity Modulation and Frequency Modulation 36
    3.5.4   RIN 38
    3.5.5   Emission Linewidth 40
3.6   Polarization Properties 40
3.7   VCSEL Applications 42

4 Design and Fabrication of VCSELs for Miniaturized Atomic Clocks   43
4.1   Adjustment of Layer Thicknesses 43
4.2   Design of the Active Region 45
    4.2.1   Bandgap Energy of Bulk AlGaAs and InGaAs 45
    4.2.2   Mechanical Strain Effect 46
    4.2.3   Bandgap Renormalization 50
    4.2.4   Relative Band Offset 50
    4.2.5   Quantum Effect 52
    4.2.6   Experimental Verification 57
4.3   Single-Mode Emission 58
4.4   Polarization Control 61
    4.4.1   Concept of Surface Gratings for Polarization Control 64
    4.4.2   Design of Surface Gratings 65
    4.4.3   Simulations of Surface Grating VCSELs 67
4.5   Layer Structure 71
4.6   VCSEL Chip Design and Processing 72
    4.6.1   Flip-Chip-Bondable Design 72
    4.6.2   VCSEL Processing 73

5 Experimental Characterization of Atomic Clock VCSELs   83
5.1   Static Characteristics 83
    5.1.1   Operation Characteristics and Emission Spectra 84
    5.1.2   Polarization Control 84
    5.1.3   Far-Field Properties 87
5.2   Dynamic Characteristics 89
    5.2.1   Small-Signal Modulation Response 89
    5.2.2   Intrinsic Modulation Behavior 90

6 Improved and Alternative Atomic Clock VCSELs   101
6.1   Modification of the Top Bragg Mirrors 101

6.2	Alternative Surface Grating Approaches	104
	6.2.1 Regular Grating VCSELs	105
	6.2.2 Inverted Grating Relief VCSELs	108
6.3	Reduction of Processing Complexity	114
6.4	Reliability Tests	118

7 Experimental Cesium-Based Atomic Clock Demonstrator — 123
7.1	VCSEL Description and Packaging	123
	7.1.1 Standard VCSELs	124
	7.1.2 Inverted Grating Relief VCSELs	126
7.2	Laser Noise and Dynamics	127
	7.2.1 Emission Linewidth	127
	7.2.2 Relative Intensity and Frequency Noise	129
	7.2.3 Modulation Sideband Characteristics	129
7.3	CPT Resonance Signal Measurement	130

8 Conclusion — 133

A Cesium Properties — 137
| A.1 | Fine and Hyperfine Structure | 137 |
| A.2 | Zeeman Splitting | 140 |

B MAC-TFC Consortium — 143

C Mask Layouts — 145

D VCSEL Processing — 147
| D.1 | Flip-Chip-Bondable VCSELs with Thick Planarization Layers | 147 |
| D.2 | Flip-Chip-Bondable VCSELs with Thin Planarization Layer (Simpler Processing) | 151 |

E VCSEL Epitaxial Structure — 153

F Experimental Measurement Setups — 155
F.1	Polarization-Resolved Operation Characteristics and Emission Spectra	155
F.2	Far-Field Measurements	157
F.3	Small-Signal Modulation Response Measurements	158
F.4	RIN Measurements	160

G Cesium Absorption Spectra — 163

H List of Acronyms   **167**

I List of Symbols   **171**
    I.1 Mathematical Operators, Special Functions and Constants 171
    I.2 Mathematical Symbols 171
    I.3 Greek Symbols 178

Bibliography   **185**

Chapter 1

Introduction and Motivation

Frequency standards or clocks provide time references for a wide range of systems and applications such as synchronization of communication networks, remote sensing and global positioning. Over the last couple of decades, demands on the data rates of many communication systems have substantially increased, imposing more restricted requirements on the stability and accuracy of their timing devices. At the same time applications have become more mobile, increasing the demand for small frequency references with low power consumption.

Atomic clocks have provided the most stable frequency references for more than 50 years [1, 2]. However, the size and power requirements of microwave-cavity-based atomic clocks prohibit them from being portable and battery-operated. Hence, research on miniaturized atomic clocks (MACs) has been carried out by various research groups. The first demonstrations of MACs were done in 2004 separately by two research groups in the United States of America, led by the National Institute of Standards and Technology (NIST) [3] and Symmetricom [4]. A Joint European research project on MACs, funded by the European commission started in 2008 [5, 6]. This dissertation reports on the achievements within the European research project in the design, fabrication and characterization of suitable laser sources for such atomic clocks. MACs use the principle of all-optical coherent population trapping (CPT) excitation which does not require a microwave cavity [7, 8]. Owing to their enhanced stability and low power consumption compared to thermally stabilized quartz-based oscillators, MACs are becoming key elements for the above-stated applications and systems. The CPT excitation is obtained in an extremely compact cesium-based vapor cell of a few cubic millimeters volume which is illuminated by an intensity-modulated laser source at a GHz-range modulation frequency.

Vertical-cavity surface-emitting lasers (VCSELs) are compelling light sources for MACs because of their low power consumption, high modulation bandwidth, and favorable beam characteristics. Similar to their use in tunable diode laser absorption spectroscopy (TD-

LAS) for regular gas sensing, VCSELs must feature strictly polarization-stable single-mode emission. Additionally, they must provide narrow linewidth emission at a center wavelength of about 894.6 nm and be well suited for harmonic modulation at about 4.6 GHz in order to employ the CPT effect at the cesium D_1 line. VCSELs emitting at 894.6 nm have been developed and employed in prototype atomic clocks [9, 10]. Those standard VCSELs with circularly symmetric resonators are often found to be polarization-stable. However, the stability cannot be guaranteed, especially after handling steps like soldering or bonding, which are necessary for microsystem integration and which might cause internal strain. Such unpredictable behavior can considerably reduce the yield of suitable devices from a fabricated wafer. In fact, owing to the cylindrical symmetry of the VCSEL resonator and the isotropic gain and reflectivity provided by the quantum wells and the Bragg mirrors, respectively, the polarization orientation of the emitted light of a standard VCSEL is a priori unknown. In the worst case, the orientation of the polarization can change during operation [11]. Therefore, several attempts have been undertaken in the past to lift the symmetry of the VCSEL structure and the isotropic property of the gain and reflectivity in the device in order to stabilize the polarization in a fixed direction [12]. Among all of these, the incorporation of a linear semiconductor surface grating at the outcoupling facet was found to be the most advantageous. The polarizing effect is induced by the difference in optical losses and thus threshold gains of modes polarized parallel or orthogonal to the grating lines.

The VCSELs for cesium-based MACs, designed and fabricated during the research performed for this dissertation, employ such pure semiconductor–air surface gratings as will be discussed thoroughly in the upcoming chapters. For the purpose of integration with the clock microsystem, flip-chip-bondable VCSEL chip designs are realized and developed. Such chip designs facilitate not only a straightforward mounting but also make the electrical contacts high-frequency compatible. Extensive static and dynamic VCSEL characterization has been performed along with several optimization cycles, supported by numerical simulations of the laser resonator.

The dissertation is organized as follows: first, general key concepts of clocks and frequency standards along with a brief historical overview on their development are introduced. Subsequently, the concept of atomic clocks, their performance, the necessity of their miniaturization, the essentiality of VCSELs as their laser sources and their main applications are discussed. The fundamentals of VCSEL design, operation and applications are presented in Chap. 3. The design of single-mode polarization-stable VCSELs emitting at 894.6 nm wavelength suitable for cesium-based MACs is outlined in Chap. 4. The concept of surface gratings for polarization control is presented in the same chapter, where the design of the gratings is based on advanced electromagnetic simulations using a fully-vectorial three-dimensional model [13]. At the end of the chapter, the layer structure of atomic clock VCSELs, the chip design and the fabrication process are discussed. Static and dynamic characteristics of initial generations of atomic clock VCSELs are discussed in Chap. 5, before further improvements and alternatives of such VCSELs are presented in Chap. 6. In the same chapter, preliminary reliability tests of some VCSELs are reported. Investigations on some atomic clock VCSELs, proving their high-level performance and their validity as laser sources for MACs, are discussed in Chap. 7. Such investigations include characterizations of noise and harmonic modulation properties of individual stand-alone VCSELs as well as delineation of the CPT signal of a prototype atomic clock employing such VCSELs. Finally, a conclusion is given in Chap. 8.

1 Introduction and Motivation

Chapter 2

Clocks and Frequency Standards: Concept, History, Miniaturization, and Applications

In this chapter, some key concepts characterizing clock performance are discussed. A brief historical overview on the development of clocks and frequency standards is given. Atomic clocks show best performance in terms of accuracy and stability. However, portable battery-operated applications require miniaturization of the atomic clock to reduce its power and size requirements, as will be explained. Finally, some applications of miniaturized atomic clocks are presented.

2.1 Key Aspects of Clocks

Clocks or frequency standards are devices which are capable of producing stable and well known frequencies with a given accuracy. They can provide necessary timing references and signals covering a wide range of frequencies which are of a great concern for vast fields in sciences and technologies [2]. A clock consists of two parts, an oscillator which produces stable periodic events and a counter which counts these events and displays the time or frequency. The oscillator itself consists of two components, in particular, a resonator which generates the periodic events and an energy source which provides energy to sustain the stability of the periodic events [14,15]. Commonly, the performance of clocks is evaluated by three figures of merit, namely accuracy, stability and the quality factor of the employed resonator.

5

2.1.1 Accuracy

Accuracy is the degree of closeness of a measured value to its definition or its ideal value. Conversely, inaccuracy of a measured value is its offset from the ideal value. Practically, the clock accuracy is measured by determining the frequency offset of the periodic events generated by the oscillator from its ideal value known as nominal frequency f_{nom}. The frequency offset can be measured in either the frequency or the time domain. A simple measurement in the frequency domain involves a frequency counter to count and display the frequency output of the clock under test. The relative frequency offset can be given by [15]

$$\bar{y} = \frac{f_{\text{meas}} - f_{\text{nom}}}{f_{\text{nom}}}, \tag{2.1}$$

where f_{meas} is the reading from the frequency counter and f_{nom} is the frequency labeled on the clock oscillator. $\bar{y}$ is known also as the relative inaccuracy. Frequency offset measurements in the time domain involve phase comparison between the outputs of the clock under test and a reference clock which has to be highly accurate, e.g., the microwave atomic clocks which will be introduced in Sec. 2.2.3. The relative frequency offset in the time domain can be given by [15]

$$\bar{y} = \frac{\Delta t}{\tau}, \tag{2.2}$$

where Δt is the amount of the time deviation and τ is the measurement period. Assuming a clock accumulates 1 μs of time deviation with respect to a reference clock over a measurement period of 24 hours (i.e., 86,400,000,000 μs), the relative frequency offset or the relative inaccuracy has thus a value of $\bar{y} = 1.16 \times 10^{-11}$.

2.1.2 Stability

Different from accuracy, which indicates how well a clock is set on its nominal frequency, stability indicates how well a clock can produce the same frequency offset over a given interval of time. In practice, clock stability is estimated by measuring frequency offsets over a given interval of time with respect to the mean frequency. In the simplest method, stability can be determined by estimating the standard deviation of a data set of measured frequency offsets. However, frequency offsets are usually a non-stationary data set, since they are time dependent. Thus the mean and the standard deviation often do not converge to particular values. Instead, the mean is alternating each time a new measured data point is added. For these reasons, a non-classical statistic method called *Allan deviation* is often utilized to estimate the fractional frequency stability as a function of averaging time τ.

The Allan deviation [15]

$$\sigma_{\bar{y}}(\tau) = \sqrt{\frac{1}{2(\bar{M}-1)} \sum_{i=1}^{\bar{M}-1} [\bar{y}_{i+1}(\tau) - \bar{y}_i(\tau)]^2} \qquad (2.3)$$

is based on differences of the adjacent $\bar{y}_i(\tau)$ values rather than on the differences from the mean value (as is the case for a "true" standard deviation), where $\bar{y}_i(\tau)$ is a set of relative frequency offsets containing $\bar{y}_1(\tau)$, $\bar{y}_2(\tau)$, $\bar{y}_3(\tau)$, ..., $\bar{y}_{\bar{M}}(\tau)$, and $\bar{M}$ is their number. From (2.2), $\bar{y}_i(\tau) = (t_{i+1} - t_i)/\tau$ and (2.3) can be thus rewritten as [15]

$$\sigma_{\bar{y}}(\tau) = \sqrt{\frac{1}{2(\bar{N}-2)\tau^2} \sum_{i=1}^{\bar{N}-2} [t_{i+2} - 2t_{i+1} + t_i]^2} \,, \qquad (2.4)$$

where t_i is a set of time measurements in the time domain containing t_1, t_2, t_3, ..., $t_{\bar{N}}$, and $\bar{N}$ is the number of data points in the t_i set which are equally spaced by τ.

2.1.3 Quality Factor

The quality or Q factor of a resonator is defined as the ratio of the resonance frequency over the linewidth of the resonance. Resonators with high resonance frequency and narrow linewidth exhibit high Q factors. Clock accuracy and stability are closely related to the quality factor of the employed resonator. Today, the unit of "second" is defined in terms of a particular resonant frequency in the cesium (Cs) atom. Therefore, if a high-Q resonator has a resonance frequency of the Cs atom, then the clock employing such a resonator will accurately generate a "second" according to its definition. A high-Q resonator has a narrow resonance linewidth which constrains the oscillator to run always at a frequency near its resonance, i.e., higher Q factor leads to better stability. However, a clock with a high-Q resonator would show good stability but poor accuracy, if the resonance frequency of its resonator is not according to the definition of "second" [14]. Another definition of the Q factor is the ratio of the stored energy in the resonator to the energy loss per each oscillation cycle. Therefore, an ideal resonator with infinity Q factor would run for ever, given a single initial push [14].

2.2 Historical Prospective of Clocks

Sundial clocks or shadow clocks are considered to be the most ancient clocks mankind got to know in the past [14]. Such clocks count and keep the track of the axial spin of the earth around its polar axis and of its rotation around the sun. Both motions were employed as oscillators for this type of clocks. As early as 3500 B.C., ancient Egyptians had built obelisks that could be used to divide the day into several divisions [16]. Obelisks also showed the longest and shortest days of the year when the shadow at noon was the shortest and longest, respectively [16]. In addition to sundial clocks, Egyptians constructed water clocks that in their simplest form consisted of a bowl which is wide at the top and narrow at the bottom, and marked from inside with horizontal "hour" ticks. The bowl was filled with water that leaks out through a small hole in the bottom [14]. Until the 14th century, Chinese, Greeks and Romans continued to rely on water and even sand clocks [14].

2.2.1 Mechanical Clocks

Water and sand clocks suffered from many problems such as ability to freeze, poor accuracy, and limitation to measure only short time intervals. By the start of the 15th century, the focus was shifted to mechanical clocks. Among the different types of mechanical clocks, the pendulum clock was considered to be the most accurate and popular at that time. The pendulum was first realized as a device by the Italian researcher Galileo Galilei who credited that it could be used as a clock resonator. However, he did not construct a workable clock before his death in the year 1642 [14]. The first working pendulum clock was acknowledged to the Dutch physicist Christian Huygens, who built it in the year 1656. The advanced models of his clock were reported to have inaccuracies smaller than 10 s per day corresponding to a relative inaccuracy of $\bar{y} \approx 10^{-4}$ using (2.2) [2, 14]. This was a dramatic improvement over the other clocks from the past. The accuracy and stability of such clocks is mainly limited by the thermal expansion of the mechanical parts including the pendulum length and the losses in the pendulum energy due to the air resistance and the clockwork. In 1721 the English inventor George Graham improved the pendulum clock by inventing a temperature-compensated pendulum known as the *mercury pendulum*, which compensates for changes in the pendulum length due to temperature variations. This has further enhanced the relative inaccuracy of the clock to be one second per day ($\bar{y} \approx 10^{-5}$) [2]. Over the century after, continuous improvements and refinements had culminated in very stable pendulum clocks like the ones manufactured by Siegmund Riefler in Germany at the end of the 19th century. His clocks achieved inaccuracies as low as 10 ms per day ($\bar{y} \approx 10^{-7}$) and became a timing standard in many astronomical observatories until the twenties of the previous century before being replaced by the Shortt clock [2]. In 1920 William H. Shortt constructed a clock with two synchronized pendulums, one of which is a master pendulum that swings as unperturbed as possible in a vacuum housing. The second is a slave pendulum which gives the master pendulum gentle pushes to maintain its motion, and also drives the hands of the clock. This allows the master pendulum to remain free from mechanical tasks that would disturb its regularity. The inaccuracies of the Shortt clocks were lower than 2 ms per day ($\bar{y} \approx 2 \times 10^{-8}$) and lower than a second per year ($\bar{y} \approx 3 \times 10^{-8}$) [2].

2.2.2 Quartz-Based Clocks

Quartz crystals were first employed in oscillators and clocks in the 1920s. Typical Q factors of quartz resonators show values between 10^5 and 10^6 [14]. The resonance of the quartz crystal is caused by the piezoelectric effect which is a mutual effect between the mechanical stress and the electric field produced by the crystal [14]. Specifically, the quartz crystal vibrates when applying an alternating electric field. At the same time, it generates an oscillating electric field because of its vibration. By employing suitable electronic circuitries, the piezoelectric effect causes crystal vibration and generates an electric signal of relatively constant frequency that can operate an electronic clock display. The first quartz-based clock was developed by the American scientist Warren A. Marrison in the year 1929, which was enclosed in a cabinet having a volume of $7.5\,\mathrm{m}^3$ [14]. For several decades quartz wrist watches are available commercially. This gives an indication of the great breakthroughs done during the last century to miniaturize the electronic circuitries.

The frequency stability of a quartz-based oscillator in the short term is influenced by fluctuations of several ambient parameters, such as temperature, humidity, pressure, external magnetic fields, external vibrations, shocks, and noise in the electronic circuitries [2,14,15]. On the other hand, the frequency stability in the long term is influenced by crystal aging [14].

Compensating the thermal frequency shift of quartz crystals has been of great interest. The first compensation attempt was made by developing temperature-compensated crystal oscillators (TCXOs) where a temperature sensor is employed to generate a correction voltage to be applied to a varactor. The varactor thus produces an equal frequency change with an opposite sign to the one originating from temperature fluctuations. Inaccuracies less than 1 minute per year ($\bar{y} \approx 2 \times 10^{-6}$), and short-term instabilities (Allan deviation) of $\sigma_{\bar{y}}(\tau) = 1 \times 10^{-9}$ over $1\,\mathrm{s}$ averaging time τ were achieved [17]. A better performance has been obtained with microcomputer-compensated crystal oscillators (MCXOs) which use a dual-mode oscillator operating simultaneously on the fundamental frequency f_1 and on the third-order harmonic (i.e., $f_3 \approx 3f_1$). Their frequency difference (i.e., $|3f_1 - f_3|$) exhibits an almost linear variation with temperature of about $-14\,\mathrm{Hz/K}$ [18]. This frequency difference is monitored by a microcomputer to correct the output frequency accordingly.

The relative inaccuracy of such oscillators is less than 1.5 s per year ($\bar{y} \approx 5 \times 10^{-8}$) while its short-term instability is about $\sigma_{\bar{y}}(\tau) = 3 \times 10^{-10}$ for $\tau = 1\,\text{s}$ [17]. A further performance improvement has been carried out by employing a very stable oven in which the crystal quartz and the other temperature-sensitive electronics are placed. Such an ensemble is known as oven-controlled crystal oscillator (OCXO). The temperature of the oven is continuously adjusted to produce a stable frequency with almost zero temperature sensitivity. A relative inaccuracy of less than 300 ms per year ($\bar{y} \approx 1 \times 10^{-8}$) and a short-term instability of $\sigma_{\bar{y}}(\tau) = 1 \times 10^{-12}$ for $\tau = 1\,\text{s}$ are achieved [17]. However, the enhanced performance of OCXOs is at the cost of large power consumption of several hundred milliwatts which is one order of magnitude larger than the power consumption by the cheaper and less accurate TCXOs [17].

2.2.3 Microwave Atomic Clocks

The next big step in the world of clocks was the use of atoms as resonators [14]. Q factors for such resonators varies between 10^5 and 10^9 [14]. In analogy to mechanical clocks, atomic clocks employ a quantum mechanical system as a "pendulum", where the oscillating frequency is related to the energy difference between two atomic quantum states. Early atomic clocks utilized cesium, namely the ^{133}Cs isotope [2]. Fundamental properties of this isotope including its energy level system are explained and illustrated in detail in App. A. Successful prototypes of Cs-based atomic clocks having a laboratory room size were developed between the years 1948 and 1955 by several research groups in the USA including the National Bureau of Standards (NBS, now National Institute of Standards and Technology, NIST), and in England at the National Physical Laboratory (NPL) [2]. They were known as *atomic cesium beam clocks*. Essen and Parry at NPL operated the first prototype of such a frequency standard and measured the hyperfine frequency splitting of the Cs ground level $6S_{1/2}$ [19]. Soon after, specifically in the year 1958, the first commercial cesium atomic beam clocks (Atomichron®) [20] became available with a relative inaccuracy of $\bar{y} = 1 \times 10^{-9}$ and a short-term instability of $\sigma_{\bar{y}}(\tau) = 5 \times 10^{-12}$ for $\tau = 100\,\text{s}$ [21]. In the following decades, a number of laboratory-sized atomic frequency standards were developed all over the world with accuracies of best clocks roughly improving by an order of magnitude per decade [2]. This development led to the re-definition of the "second" in the year 1967 when the 13th General Conference on Weights and Measures (CGPM) defined the "second" as *the duration of 9 192 631 770 periods of the radiation corresponding to the transition between the two hyperfine levels of the ground state of the cesium 133 atom*. This definition was adopted by the International System of Units (SI). Two decades after, the relative inaccuracies of atomic cesium beam clocks (e.g., CS2

in 1986 at the Physikalisch-Technische Bundesanstalt (PTB), Germany) were as low as $\bar{y} = 2.2 \times 10^{-14}$ with a short-term instability of $\sigma_{\bar{y}}(\tau) = 2.7 \times 10^{-12}$ for $\tau = 1\,\text{s}$ [22].
A new era of cesium atomic clocks started when a prototype of an *atomic cesium fountain clock* was set up at the Laboratoire Primaire du Temps et des Fréquences (LPTF)[1] in Paris [23]. In such type of atomic clocks, Cs atoms are laser cooled and follow a ballistic flight in a gravitational field for about one second. The long interaction time made possible by the laser cooling method leads to a reduced linewidth of the resonance curve. Less than two decades after the first implementation, the relative inaccuracy of fountain clocks has become less than 1×10^{-15} [2]. In particular, relative inaccuracies of 7.2×10^{-16}, 2.5×10^{-16} and 3.2×10^{-16} have been reached by the atomic cesium fountain clocks: FO1 at BNM-SYRTE in France [24], CSF1 at PTB in Germany [25] and NIST-F1 at NIST in USA [26–28], respectively. Frequency instabilities of the three atomic clocks are $\sigma_{\bar{y}}(\tau) = 2.9 \times 10^{-14}(\tau/\text{s})^{-0.5}$ for $\tau = 5$ to $60\,\text{s}$, $\sigma_{\bar{y}}(\tau) = 7.4 \times 10^{-14}(\tau/\text{s})^{-0.5}$ for $\tau = 100$ to $3000\,\text{s}$, and $\sigma_{\bar{y}}(\tau) = 2.8 \times 10^{-13}(\tau/\text{s})^{-0.5}$ for $\tau = 4$ to $4000\,\text{s}$, respectively [24–28].

2.3 Miniaturization of Atomic Clocks

In comparison to quartz-based clocks, laboratory-sized atomic frequency standards (e.g., atomic cesium beam and atomic cesium fountain) provide much better frequency accuracies and stabilities as stated in the previous section. However, their very large sizes (e.g., few cubic meters), costs (e.g., several hundreds thousands of US dollars) and power consumptions (e.g., several hundreds of watts) restrict their use in real-world applications [29]. Smaller versions of optically pumped rubidium atomic frequency standards (RAFS) [30, 31] with volumes around $200\,\text{cm}^3$ and power consumptions of a few watts have become commercial standards. Relative inaccuracies of less than $\bar{y} = 3 \times 10^{-13}$ and short-term instabilities of $\sigma_{\bar{y}}(\tau) = 3 \times 10^{-11}(\tau/\text{s})^{-0.5}$ for $\tau = 1$ to $1000\,\text{s}$ were reported for such atomic standards [30]. The vapor cell is filled with ^{87}Rb and surrounded by a microwave cavity which is in the centimeter scale. Optical pumping in RAFS is done with a rubidium discharge lamp which is spectrally filtered by a ^{85}Rb filter cell [29]. However, a large number of portable in-field applications require smaller and less expensive frequency references with much lower power consumptions. TCXOs are low-power low-cost small-sized frequency references useful in several battery-operated portable applications which require moderate performance in terms of frequency accuracy and stability. Nevertheless,

---
[1]Later it became the Bureau National de Métrologie - Systèmes de Référence Temps-Espace, BNM-SYRTE, and now the Laboratoire National de Métrologie et d'Essais - Systèmes de Référence Temps-Espace, LNE-SYRTE.

their frequency stability at longer averaging times (one hour to several days) is not sufficient for the requirements of many applications in the civil and military navigation and communication sectors [29]. The size of RAFS is mainly determined by the size of the microwave cavity, and the most of power is consumed in heating the rubidium discharge lamp [29]. Therefore, in order to reduce its power consumption and to miniaturize its size, it is necessary to replace the lamp and the filter cell with a low-power laser source and to eliminate the microwave cavity. The only way to achieve such volume and power reduction is to move to a clock concept which does not rely on such cavities and lamps. A novel method based on the so-called all-optical CPT spectroscopy has been proposed [7,8]. CPT excitation is obtained in an extremely compact microfabricated alkali vapor cell of a few cubic millimeters volume which is illuminated by a single-mode VCSEL under harmonic modulation. CPT spectroscopy will be explained in some detail in Sect. 2.3.1. In addition to employing CPT, realization of MACs[2] requires fabrication techniques developed for microelectromechanical systems (MEMS). Such techniques are mainly applied to microfabricate the alkali vapor cells, as will be described in Sect. 2.3.2. By the CPT technique and MEMS technology, MACs can become the most attractive timing solution among the existing technologies, since it could provide: i) superior stability at medium and long-term observation times compared to existing low-cost quartz-based oscillators, ii) compact sizes and significantly low manufacturing costs in comparison to the existing RAFS modules and iii) lower power consumptions which would significantly simplify thermal management issues.

Research on MACs has been carried out by various research groups. The first demonstrations of MACs based on CPT spectroscopy and MEMS fabrication techniques were done in 2004 collaboratively by two research groups in the United States of America, led by NIST [3] and Symmetricom [4]. Such frequency sources have recently become commercially available [32,33]. In 2008, the European commission started to fund a collaborative research project within its seventh framework programme (FP7) for research and technological development to realize the first European MAC. It was called *MEMS atomic clocks for timing, frequency control & communications* (MAC-TFC) [5,6]. The objective of MAC-TFC was to develop and demonstrate all necessary technology to achieve a miniaturized battery-operated atomic clock having a volume less than $10\,\text{cm}^3$, a power consumption not exceeding $155\,\text{mW}$, and a short-term instability of $\sigma_{\bar{y}}(\tau) = 5 \times 10^{-11}$ over one hour averaging time [5,6]. In general, MACs can utilize the same vapors as microwave atomic clocks, namely rubidium (Rb) or cesium (Cs). They require laser wavelengths of $780.2\,\text{nm}$ (Rb D_2 transition line), $795.0\,\text{nm}$ (Rb D_1), $852.4\,\text{nm}$ (Cs D_2), or $894.6\,\text{nm}$ (Cs

---

[2]As mentioned in Chap. 1, MAC stands for miniaturized atomic clock, or alternatively MEMS atomic clock.

2.3 Miniaturization of Atomic Clocks

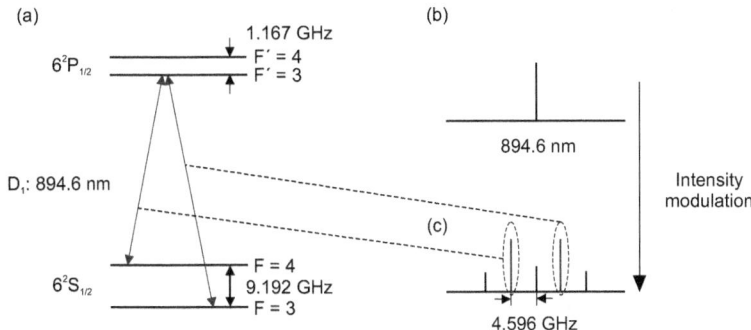

Figure 2.1: Simplified Cs energy level diagram showing the D_1 optical transition and the hyperfine splitting of the ground state $6^2S_{1/2}$ and the excited state $6^2P_{1/2}$ (a). Illustration of the optical spectra of a CW-operated VCSEL (b) and of an intensity-modulated VCSEL (c).

D_1) [9]. It has been shown that excitation on the D_1 line results in higher contrast and narrower CPT resonance signals[3] compared to the D_2 line, for both Rb [34] and Cs [35]. For MAC-TFC, the Cs D_1 line was employed.

The MAC-TFC consortium, including 10 partners from academic, research and industrial sectors, possessed all required knowledge and technologies in order to achieve the MAC-TFC objectives. Names and locations of the consortium partners can be found in App. B.

2.3.1 Three-Level System and CPT Spectroscopy

CPT excitation is based on the quantum mechanical phenomenon that the population probability of the highest energy level in a three-level system (also known as a Λ system) can be drastically reduced via illumination by two coherent light sources whose emission wavelengths match the transition energies between the two lower energy levels and the upper level [7, 36]. Figure 2.1 (a) depicts a simplified Cs energy level diagram showing the D_1 line. Due to the interaction between the total angular momentum of the electrons and total angular momentum of the nuclear spin, the ground level $6^2S_{1/2}$ and the excited level $6^2P_{1/2}$ are split into two hyperfine levels separated by about 9.192 GHz and 1.167 GHz frequency differences, respectively. The hyperfine levels are identified by the quantum number F associated with the total atomic angular momentum **F**. As explained in App. A.1, to distinguish between the hyperfine structures of $6^2S_{1/2}$ and $6^2P_{1/2}$ levels,

---
[3] A typical CPT resonance signal is visualized in Fig. 2.2.

13

2 Clocks and Frequency Standards: Concept, History, Miniaturization, and Applications

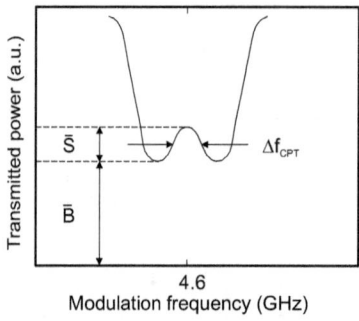

Figure 2.2: Typical CPT curve showing the transmitted optical power through a Cs vapor cell as a function of the VCSEL modulation frequency, where the VCSEL wavelength is 894.6 nm.

they are assigned quantum numbers F and F', respectively, and take values of 3 and 4, as depicted in Fig. 2.1 (a). More information on the fine and hyperfine structure of the Cs energy system can be found in App. A.1. Two separate lasers can be used to achieve the CPT effect. However, since the two ground energy levels are very close, it is also possible to produce light with the two required wavelengths by modulating a single laser source with a radio frequency (RF) signal at a frequency that is equal to half of the Cs hyperfine ground splitting frequency. One then makes use of the modulation sidebands.

VCSELs are ideal laser sources for this purpose owing to high-speed, high-efficiency modulation capability and low power consumption. Figures 2.1 (b) and (c) show schematic emission spectra of a VCSEL operating in continuous-wave (CW) mode and of an intensity-modulated VCSEL at 4.596 GHz modulation frequency, respectively. The RF power is distributed over several modulation sidebands which are equally separated by the modulation frequency. The first-order sidebands at ±4.596 GHz can be employed for the operation of the atomic clock. Vapor of Cs atoms in the CPT condition shows reduced absorption or increased transparency at 4.596 GHz modulation frequency, as depicted in Fig. 2.2. Therefore this phenomenon is also called electromagnetically induced transparency (EIT) [37]. The contrast of the CPT signal is defined by the ratio of the signal amplitude $\bar{S}$ and the direct current (DC) background noise $\bar{B}$ as [9]

$$\bar{C} = \frac{\bar{S}}{\bar{B}}, \tag{2.5}$$

where the background noise arises from the photodetector noise and the laser intensity and frequency noise. CPT signals with high contrast $\bar{S}$ and narrow linewidth[4] Δf_{CPT} are highly desirable to achieve reduced short-term instabilities for CPT-based atomic clocks [29].

---
[4] The linewidth of any signal over the entire dissertation is defined to be the full width of the signal at half maximum (FWHM).

2.3 Miniaturization of Atomic Clocks

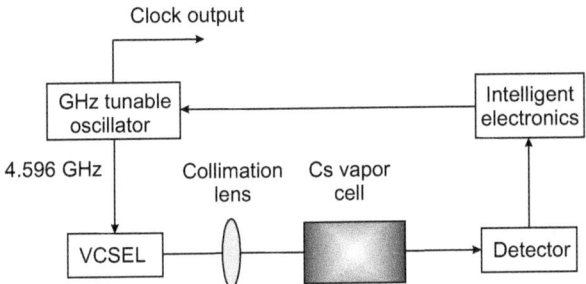

Figure 2.3: Simplified block diagram of a Cs-based MAC.

A simple block diagram of a MAC is depicted in Fig. 2.3. It comprises a VCSEL, a Cs vapor cell, a photodetector, a tunable oscillator, optics and "intelligent" electronics. The simplest clock configuration is to have the laser beam, the vapor cell and the photodetector, all in a single straight line. The photodetector receives the laser light transmitted through the vapor cell. The oscillator modulates the laser current at a frequency of about 4.596 GHz. As mentioned above, the first-order sidebands are separated by 9.192 GHz, which is equal to the Cs hyperfine splitting frequency of the two ground levels (i.e., $F = 3$ and $F = 4$), and can be thus employed for CPT excitation. A CPT signal similar to the one depicted in Fig. 2.2 is detected and used as a timekeeping element for the clock. The "intelligent" electronics search for and detect the modulation frequency achieving a maximum amplitude of the CPT signal. At this frequency, the tunable oscillator has to be locked and precisely tuned. By this means a stable clock output is obtained. It is worth to mention that the block diagrams of real MACs are much more complicated than the one shown in Fig. 2.3. They include several servo loops (or servos) to adjust and stabilize important parameters such as VCSEL temperature and bias current[5], vapor cell temperature, RF modulation power and frequency, and other parameters. Such servos are operated at low frequencies ranging from a few tens of Hertz to a few kHz.

2.3.2 Microfabricated Alkali Vapor Cells

The heart component of a MAC is a microfabricated cell containing alkali vapor and a buffer gas atmosphere. Recently, FEMTO-ST[6] has successfully microfabricated alkali vapor cells based on an extremely compact sealed cavity. The cell is formed by etching a

---
[5]By these parameters the VCSEL output power and frequency are stabilized.
[6]FEMTO-ST is a MAC-TFC project partner. See App. B to know more about the MAC-TFC consortium.

15

Figure 2.4: Microfabricated alkali vapor cells on a one-cent Euro coin [6].

silicon wafer and sealing it between two borosilicate glass wafers using anodic bonding in an atmosphere of a buffer gas (or gases) [38, 39]. The cell vapor cavity has an optical window with 2 mm diameter and 1.4 mm length, as depicted in Fig. 2.4. After fully fabricating the cell, Cs vapor is generated by local heating of a side-cavity (of 1.65 mm × 1.65 mm size) containing Cs metallic dispenser using a high-power infrared laser source [38, 39]. Reliability tests of the microfabricated cells have demonstrated excellent long-term stability. A major contribution of the CPT linewidth are the collisions of the alkali atoms with the cell walls [29]. To reduce such an effect, the vapor cell is filled with an inert buffer gas (e.g., Ne or Ar) or molecules (e.g., N_2), thus the time between wall collisions can be prolonged and the residual Doppler broadening of the CPT signal can be reduced [40, 41].

2.3.3 Frequency Stability of MACs

For most applications of MACs the absolute clock frequency is of minor interest, since it can be calibrated. However, it is of great importance that the frequency does not change over time, i.e., it needs to have good stability, but not necessarily good accuracy [29]. Over short averaging times τ, the stability of many atomic clocks including MACs is characterized by white frequency noise, where $\sigma_{\bar{y}}(\tau)$ can be expressed as [35]

$$\sigma_{\bar{y}}(\tau) \propto \frac{\Delta f_{\text{CPT}} \bar{N}_{\text{MAC}}}{\bar{S}} \tau^{-0.5}, \qquad (2.6)$$

with $\bar{N}_{\text{MAC}}$ being the overall noise of the MAC at the RF modulation signal. There are many contributions of $\bar{N}_{\text{MAC}}$ including shot and thermal noise of the photodetector, laser frequency and intensity noise, phase noise of the tunable oscillator and noise of the electronics. The proportionality of $\sigma_{\bar{y}}(\tau)$ to $\tau^{-0.5}$ is commonly found in practice for short averaging times τ and is almost a standard way to represent $\sigma_{\bar{y}}(\tau)$, as mentioned at the end of Sect. 2.2.3. According to (2.6), it is desirable to have a high signal-to-noise ratio $\bar{S}/\bar{N}_{\text{MAC}}$ and a narrow linewidth Δf_{CPT} of the CPT signal in order to achieve low

2.3 Miniaturization of Atomic Clocks

short-term instabilities. The frequency noise of VCSELs is known to be one of the main limitations of the short-term stability of MACs [35]. Frequency-to-amplitude conversion noise, which originates from the sensitivity of the atomic absorption resonances of the alkali vapor to the laser phase fluctuations, contributes to $\bar{N}_{MAC}$ [42]. It is thus advantageous to employ a buffer gas with high pressure by which the atomic resonance lines become broader[7] and exhibit thus a less steep frequency-to-amplitude conversion slope [29]. Another laser noise, which could distort the CPT signal (i.e., reduce its contrast and increase its linewidth), is the mode competition noise between the different polarization modes [44]. Such noise would thus degrade the short-term stability of MACs. In the worst case, a CPT signal could completely vanish due to the shift in the laser emission frequency associated with polarization switches. The difference between the emission frequencies of the two VCSEL polarization modes could be up to 80 GHz [45] which is at least four times the hyperfine splitting frequency between the Cs ground levels. Therefore, the spectral overlap of the atomic resonance lines and the laser modulation sidebands would be definitely lost. Avoiding polarization switches during the VCSEL operation is of a great interest for MACs[8].

Over long averaging times, the frequency stability of MACs is limited by frequency shifts and asymmetries of the atomic resonance lines. They can originate from perturbations or fluctuations of environmental magnetic fields, buffer gas pressure, temperature, acceleration, RF modulation power or light shift [29]. Therefore, it is highly critical to control and stabilize these parameters. Light shift refers to a frequency shift of the atomic resonance lines via the *dynamic Stark effect* [46]. It originates from interaction between the alkali atoms and the electromagnetic field of the laser emission. The shift depends on both the frequency and intensity of the light. Therefore, lasers with low frequency noise and relative intensity noise (RIN) characteristics are indispensable to achieve highly stable MACs. For CPT excitation only the two first-order sidebands of an intensity-modulated VCSEL are employed. However, all the sidebands (including the carrier, see Fig. 2.1 (c)) contribute to the light shift. As the VCSEL ages, its operating temperature and current, resonant to the D_1 wavelength, change. Additionally, the VCSEL modulation characteristics (which is a function of its operating temperature and current) change as well. Consequently, fluctuations of the laser output power and frequency take place and result in a drift of the clock output frequency via the light shift effect [47]. This limits the long-term stability of the clock. A proper selection of the RF modulation power could cancel

---

[7]The higher the buffer gas pressure in an alkali vapor cell, the broader are the atomic resonance lines due to the collisional-broadening effect [43].

[8]Properties and stabilization of polarization modes in VCSELs are of great concern for this dissertation. More discussions on polarization properties as well as polarization control will be presented in Sect. 3.6 and Sect. 4.4, respectively.

the total light shift caused by the laser carrier and all other sidebands, as demonstrated for Cs microfabricated vapor cells filled by Ne buffer gas [48]. This result highlights the fact that improving the long-term stability of MACs requires low noise and stable RF modulation power. The RF level could potentially be locked to this point, where no light shifts are present, via an additional servo loop [49]. As mentioned in Sect. 2.3.2, microfabricated vapor cells are usually filled with buffer gas to narrow the CPT linewidth. This improves the short-term frequency stability according to (2.6). However, the buffer gas changes the hyperfine splitting frequency and the linewidth of the atomic resonance lines of the alkali vapor [43]. Therefore, it is extremely important that the pressure of the buffer gas remains stable over a long aging time, since pressure fluctuations induce frequency variations and consequently affect the long-term stability of the clock. Such frequency variations are usually temperature-dependent. For RAFS, it is common to use a mixture of two buffer gases which induces frequency shifts with opposite signs and, therefore, the thermal sensitivity of the clock frequency could be set to zero at a certain cell temperature T_{inv} commonly known as *inversion temperature* [50]. In particular, the clock frequency shows a quadratic dependence on cell temperature around T_{inv}. However, the partial and total buffer gas pressures are very difficult to control in microfabricated vapor cells because of the high temperatures required for the anodic bonding process, which could create residual gases inside the cell [29]. Recently FEMTO-ST has reported such a quadratic dependence around $T_{\text{inv}} = 80°C$ using only a single buffer gas, namely Ne [51]. Employing a single buffer gas instead of two buffer gases strongly relaxes constraints on the accuracy control of the gas pressure during the cell filling procedure. Another favorable aspect for the long-term frequency stability is that T_{inv} for Ne is found to be independent of its pressure [48].

2.3.4 Requirements on the Laser Source

Over the last few years, MACs have emerged as a new application field of VCSELs. They are compelling light sources for such clocks due to their low power consumption, high modulation bandwidth, and favorable beam characteristics. Their sub-mA threshold currents are favorable for small power consumption, and hybrid integration with the clock microsystem is straightforward. VCSELs must feature strictly polarization-stable single-mode emission. Additionally, they must provide low-noise narrow linewidth emission at a center wavelength of about 894.6 nm and be well suited for harmonic modulation at about 4.596 GHz with 65 to 80°C operating temperature in order to employ the CPT effect at the Cs D_1 line. Table 2.1 states main device parameters and their target values as required for use in MACs according to the MAC-TFC goals. Frequencies for the target

Table 2.1: Requirements on VCSEL parameters for MAC-TFC.

Parameter	Value
Operating temperature	65–80 °C
Threshold current	< 1 mA
Optical power	≈ 100 µW
Emission wavelength	894.6 nm
Suppression ratio (for transverse and polarization mode)	> 20 dB
3 dB modulation bandwidth	> 5 GHz
RIN	< −120 dB/Hz at 1 kHz and < −102 dB/Hz at 500 Hz
Emission linewidth	< 100 MHz

values of the relative intensity noise are the ones at which servo loops are operated.

2.4 Applications of MACs

Many recent civil and military applications are increasingly demanding battery-operated small-sized frequency references that provide highly accurate and stable timing signals. Owing to their enhanced stability compared to TCXOs and much lower power consumption and smaller size compared to OCXOs and RAFS, MACs are key elements in a wide range of portable systems and applications such as satellite-based navigation systems (SBNS), synchronization of mobile communication networks, or undersea exploration.

For example, MAC-based SBNS receivers could lock more rapidly to satellites, allowing for faster acquisition of precise positioning information [52]. It also becomes possible to operate with only three in-view SBNS satellites for long hold-over time periods (e.g., > 24 hours) without the necessity of the fourth SBNS satellite which provides the synchronization signal [33, 53]. With no doubt such a function property is, e.g., highly critical for military navigation and tracking systems specially in urban locations [53].

MACs would be also important elements for the precision time protocol (PTP), which is an approach to distribute synchronization over IP-based packet networks (e.g., Ethernet[9]) [54]. The PTP algorithm calculates the time difference between network nodes and then

---
[9]Also known as IEEE 1588 protocol which aims to achieve sub-µs real-time synchronization of clocks in networked distributed measurement, automation and control systems which are mainly installed in industrial environments or power distribution networks.

updates the time, phase and frequency at the slave nodes to match the master clocks. PTP network topology consists of a hierarchy of master–slave clocks. Master clocks are typically SBNS-based timing devices while slave clocks are typically OCXOs or TCXOs [55]. For their unique features of enhanced frequency stability while having small size, weight, and power (SWaP) [33], MACs would be the best replacement of OCXOs or TCXOs for portable low-power applications of IP-based networks.

Undersea sensors, mainly used for seismic research, and oil or natural gas exploration, require very stable clocks to time-stamp the data collected by the sensor. Such sensors lie on the ocean floor where SBNS signals cannot reach. Historically, people used OCXOs as timing devices for these sensors. Power consumption is critical since the batteries that power these sensors have to last for the duration of the exploration mission. Power consumption of MACs is between 10 and 20% compared to OCXOs [33]. This results in smaller batteries or extended mission durations. Moreover, MACs would exhibit two orders of magnitude better long-term stabilities (e.g., $\sigma_{\bar{y}}(\tau) < 3 \times 10^{-10}$ over 1 month) compared to OCXOs leading to greatly reduced time-stamping drift errors [33]. These are very important features for such SBNS-denied environments.

In-field private radio communications systems, e.g., professional mobile radio (PMR) systems, mainly used by police forces and fire brigades, are increasingly demanding an enhanced level of synchronization between the base stations [56]. Beside fixed base stations, PMR networks are based on mobile base stations where MACs could be utilized, offering much enhanced frequency stability than TCXOs and longer battery lifetime than OCXOs.

Chapter 3

Fundamentals of VCSELs

In this chapter, the major principles of VCSEL design and operation are described along with a brief discussion of the advantages of such lasers in comparison with edge-emitting lasers. Experimental results illustrating the operation and emission characteristics of standard oxide-confined VCSELs are shown. The temperature behavior in terms of the thermal resistance is discussed. The basic treatment of laser dynamics and noise behavior is introduced in terms of the small-signal modulation response and the relative intensity noise. Finally, some important applications of VCSELs are presented. A deeper theoretical introduction and description of VCSELs can be found in Refs. [57–59].

3.1 Device Structure and Properties

The VCSEL was first suggested by Iga *et al.* in 1977 [60], and first demonstrated by Soda *et al.* in 1979 [61]. Unlike edge-emitting lasers (EELs), the emission of VCSELs is orthogonal to the semiconductor wafer surface. This can be seen in Fig. 3.1, which schematically shows a cross-section of a typical top-emitting VCSEL structure. In comparison to EELs, VCSELs have several unique features such as low power consumption, wafer-level testing, favorable integration properties, high-frequency modulation capability, good temperature stability, low-cost production, and slowly divergent circular beams. The latter property allows coupling to optical fibers using low-numerical-aperture lenses with coupling efficiencies of more than 70%. An additional advantage of VCSELs is their low sensitivity to optical feedback due to the high mirror reflectivity [62]. Typical semiconductors on which VCSELs are based are GaAs and InP. All VCSELs presented in this dissertation are fabricated using the AlInGaAs material system on GaAs substrates.

The inner cavity shown in Fig. 3.1 is surrounded by two doped AlGaAs layer stacks forming the laser mirrors. The active region is composed of a few InGaAs quantum wells (QWs) embedded in AlGaAs barrier layers with some ten nm thickness. The QWs serve as

21

3 Fundamentals of VCSELs

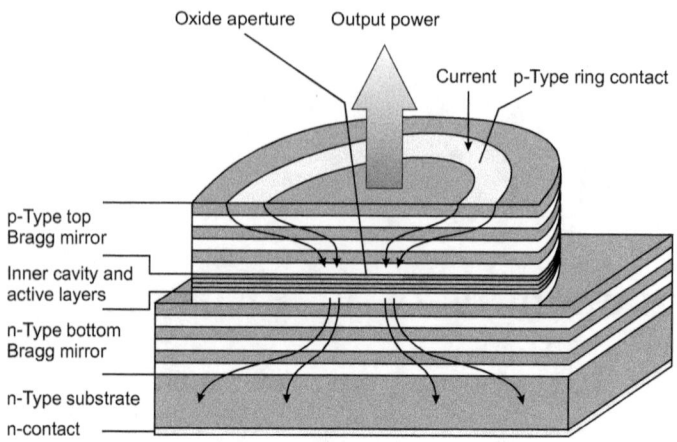

Figure 3.1: Schematic cross-section of a typical top-emitting, oxide-confined VCSEL structure.

optical amplifying layers while the laser mirrors provide optical feedback. Electric current is injected through ohmic contacts on the epitaxial and substrate side of the device. For top emission (see Fig. 3.1), the shape of the top ohmic contact has to be annular, while for bottom emission, a transparent substrate or substrate removal is required [63]. In VCSELs the length over which the light is amplified is much shorter (e.g., some ten nm) than in EELs (e.g., some hundred µm). In order to provide small threshold gain, the mirror losses have to be significantly reduced by using highly reflecting mirrors. Therefore the mirrors in VCSELs are realized as distributed Bragg reflectors (DBRs) [64,65] instead of cleaved mirrors as in conventional EELs. DBR consists of an alternating sequence of high and low refractive index layers formed by alternating layers of AlGaAs with low and high Al content, respectively. The electromagnetic wave is partially reflected at each interface. At each transition from low to high refractive index, a phase shift of π occurs. By choosing a quarter-wave thickness for each DBR layer, a standing-wave pattern is formed, as can be seen in Fig. 3.2 (left), which depicts the amplitude of the electric field $\bar{E}(z)$ and the refractive index profile $\bar{n}(z)$ in the inner part of a VCSEL designed for an emission wavelength of $\lambda = 894.6\,\mathrm{nm}$. The calculation of the standing-wave pattern is done with the one-dimensional transfer-matrix method [66] using the REFLEX code [67]. The QWs are placed in an antinode of the standing-wave pattern in order to achieve a good coupling between electrons and photons. This requires to make the inner cavity surrounding the QWs one material wavelength thick. Therefore, the inner cavity is further extended by sandwiching the QWs between two larger bandgap AlGaAs cladding layers which additionally improve the carrier confinement in the active region. A spectral feature

3.1 Device Structure and Properties

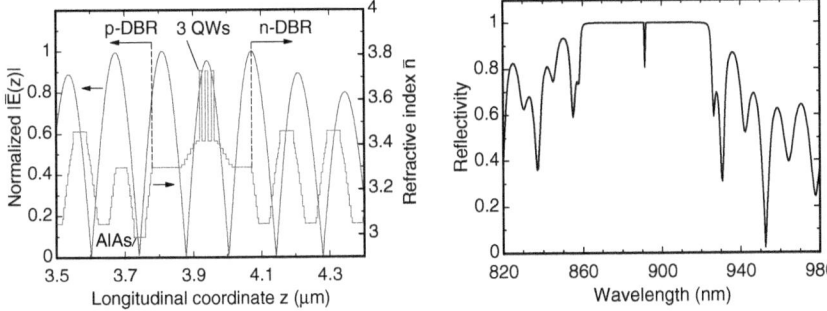

Figure 3.2: Spatial distribution of the normalized electric field amplitude in the center region of an 894.6 nm wavelength VCSEL together with the refractive index profile (left). Reflection spectrum of the same VCSEL structure which has 25 top and 38.5 bottom DBR pairs (right). Both figures are calculated using the transfer-matrix method [66, 67].

representing the resulting longitudinal mode in the inner cavity is known as the *cavity resonance*. Due to the short cavity length, modes of higher or lower order have significant spectral splitting of about 100 nm and are thus neither amplified by the QWs nor reflected by the DBRs due to the spectral limitations of the gain bandwidth of the QW material and of the stop-band of the DBR (both < 100 nm) [66]. Therefore, a VCSEL is inherently longitudinal single-mode and its emission spectrum is not determined by the gain peak as in EELs, but rather by the spectral position of the cavity resonance and by transverse modes. Figure 3.2 (right) depicts a calculated reflectivity spectrum of a full layer structure of an 894.6 nm wavelength VCSEL using the transfer-matrix method. The spectrum is a superposition of the reflection spectra of the top and bottom DBRs while the dip in the center of the stop-band signifies the presence of a single cavity resonance.

Optical confinement in the active region is described by the confinement factor

$$\Gamma = \Gamma_t \cdot \Gamma_z = \frac{V_a}{V_p}, \quad (3.1)$$

which specifies the ratio between the volumes of the active region V_a and of the photons or cavity mode V_p. It is split into a longitudinal Γ_z and transverse Γ_t part, where the latter can be approximated by $\Gamma_t \approx 1$. The longitudinal confinement factor

$$\Gamma_z = \frac{d_a}{L_{c,\text{eff}}} \quad (3.2)$$

is the ratio between the total thickness of the active layers d_a and the effective cavity length $L_{c,\text{eff}}$ which contains the one-wave thick inner cavity L_c and the effective penetration

23

3 Fundamentals of VCSELs

depths $l_{\text{eff,t}}$ and $l_{\text{eff,b}}$ into the top and bottom DBRs, respectively, and can be written as

$$L_{c,\text{eff}} = L_c + l_{\text{eff,t}} + l_{\text{eff,b}}. \tag{3.3}$$

For a simple binary quarter-wave stack DBR, the penetration depth is approximated as [66]

$$l_{\text{eff}} \approx \frac{\lambda_B}{4\Delta \bar{n}_B}, \tag{3.4}$$

where λ_B is the Bragg wavelength and $\Delta \bar{n}_B$ is the refractive index step in the binary DBR. For the VCSEL of Fig. 3.2, values of $\lambda_B = 894.6\,\text{nm}$ and $\Delta \bar{n}_B = 0.33$ lead to an $l_{\text{eff}} \approx 0.675\,\mu\text{m}$ assuming a binary DBR. The resulting $L_{c,\text{eff}}$ is approximately $1.64\,\mu\text{m}$. With three 8 nm thick QWs, $d_a = 24\,\text{nm}$ and $\Gamma = 1.46\%$ is calculated using (3.1) and (3.2) assuming $\Gamma_t = 1$.

Unlike a conventional EEL cavity, in a VCSEL cavity the active region does not extend over the full cavity length L_c but is enclosed by larger bandgap layers to form a double-heterostructure. Therefore, in order to obtain the average gain in the cavity, the spatial overlap of the standing-wave pattern $\bar{E}(z)$ with the QWs is defined as the *relative confinement factor* or the *gain enhancement factor* [66]

$$\Gamma_r = \frac{L_c \int_{d_a} |\bar{E}(z)|^2 dz}{d_a \int_{L_c} |\bar{E}(z)|^2 dz}. \tag{3.5}$$

In the general case of M_a active layers with equal gain, located at positions $z_{il} \leq z \leq z_{ih}$ with $i = 1, \ldots, M_a$ we find [66]

$$\Gamma_r = 1 + \frac{\lambda}{4\pi \langle \bar{n} \rangle} \frac{\sum_{i=1}^{M_a} \sin(4\pi \langle \bar{n} \rangle z_{ih}/\lambda) - \sin(4\pi \langle \bar{n} \rangle z_{il}/\lambda)}{\sum_{i=1}^{M_a} z_{ih} - z_{il}}, \tag{3.6}$$

where $z = 0$ is centered in the inner cavity and $\langle \bar{n} \rangle$ is the spatial average of the refractive index. For the VCSEL of Fig. 3.2 with three centered 8 nm thick QWs separated by 10 nm barriers, we get $\Gamma_r = 1.8$. Therefore, by exploiting the standing-wave effect, the available amount of optical amplification can be almost doubled.

In order to achieve current confinement to a predefined active area, selective lateral oxidation [68] of a some ten nm thick semiconductor layer directly above the inner cavity with high Al content is performed. Lateral oxidation transforms this layer to an electrically insulating material, hence the current flow is confined to the remaining aperture at the center of the layer, as illustrated in Fig. 3.1. In addition to its insulating property, the oxidized region has a much lower refractive index than the semiconductor (1.6 compared to 3) [69, 70]. This local difference in refractive index between the annular oxide section and the active device center leads to an optical waveguiding caused by a center region with high and an outer section with low average refractive index (spatially averaged over the entire stack of layers forming the VCSEL structure). The optical waveguiding can be controlled by adjusting the thickness and the longitudinal position of the oxide layer. The thinner and closer to a node of the standing-wave pattern, the lower the resulting waveguiding. The optical waveguiding has to be weakened to avoid multiple transverse mode emission. To achieve single transverse mode emission with relatively reliable oxide aperture sizes, as required in many applications like optical sensing and MACs, the oxide layer is placed at the position of the first node in the standing-wave pattern above the inner cavity, as depicted in Fig. 3.2 (left) [71, 72]. By this means, decreased overlap of the oxide layer and the electric field intensity is obtained, which weakens thus the optical waveguiding. Typical oxide aperture sizes for single transverse mode VCSEL oscillation are between 3 and 4 µm. Throughout the entire dissertation, all discussed devices have an oxidation layer consisting of pure AlAs with no Ga content.

3.2 Threshold Conditions

When a VCSEL is forward biased, carriers are injected into QWs, leading to the so-called *population inversion*, and *stimulated emission* can thus take place for sufficiently high carrier densities [73]. Stimulated emission produces coherent photons with the same energy, phase and propagation direction as the exciting wave. Therefore, *optical amplification* or *gain* is earned. For lasing, the round-trip optical gain has to balance the round-trip optical losses of the VCSEL cavity. Optical losses include mainly internal losses and light outcoupling losses. The internal losses are due to free carriers, fundamental absorption and scattering at interfaces. Outcoupling losses are due to some fraction of radiation transmitted through the DBRs. This condition of self-stimulation can be achieved when the electric field reproduces itself exactly in phase and amplitude after one complete round-trip through the effective length of the VCSEL cavity and can be written as

$$\sqrt{R_t R_b} \exp\{-2\mathrm{i}(\bar{\gamma} L_{c,\mathrm{eff}} + \phi_t + \phi_b)\} = 1, \qquad (3.7)$$

where R_t and R_b are the intensity reflection coefficients of the top and bottom DBRs, respectively, and $\mathrm{i} = \sqrt{-1}$ is the imaginary number. ϕ_t and ϕ_b are the phase shifts of the electric field at the interfaces of the inner cavity to top DBR and the inner cavity to bottom DBR, respectively. The complex propagation constant

$$\bar{\gamma} = \frac{2\pi}{\lambda} \langle \bar{n} \rangle - \mathrm{i}\frac{\alpha}{2} \qquad (3.8)$$

can be defined as a function of the intensity attenuation coefficient α and the spatial average of the refractive index $\langle \bar{n} \rangle$. With optical gain g and intrinsic losses α_i and α_a in the passive and the active sections, respectively, the intensity attenuation coefficient α can be written as

$$\alpha = \Gamma\Gamma_r(\alpha_a - g) + (1 - \Gamma)\alpha_i. \tag{3.9}$$

Replacing $\bar{\gamma}$ in (3.7) and comparing the magnitude leads to the *amplitude condition*

$$\sqrt{R_t R_b}\exp\{-\alpha L_{c,\text{eff}}\} = 1, \tag{3.10}$$

from which the threshold gain is given by

$$g_{\text{th}} = \alpha_a + \frac{1}{\Gamma_r d_a}\left[\alpha_i(L_{c,\text{eff}} - d_a) + \ln\frac{1}{\sqrt{R_t R_b}}\right], \tag{3.11}$$

using (3.1) and (3.2) and assuming $\Gamma_t = 1$. More precisely, α_i should be the average of the spatially varying absorption coefficient weighted by the standing-wave pattern of the electric field intensity $|\bar{E}(z)|^2$. Therefore, it is more convenient to determine the threshold gain g_{th} using the transfer-matrix method. R_t and R_b assume lossless top and bottom mirrors, respectively. However, for DBRs with small losses α_i, the maximum reflectivity

$$R_\alpha \approx R\exp\{-2\alpha_i l_{\text{eff}}\} \approx R(1 - 2\alpha_i l_{\text{eff}}) \tag{3.12}$$

can be defined [57], where $2\alpha_i l_{\text{eff}} \ll 1$ and assuming that the wave travels the effective penetration depth l_{eff} through the DBR forward and backward. Using (3.3), the threshold gain from (3.11) can be rewritten as [66]

$$g_{\text{th}} = \alpha_a + \frac{1}{\Gamma_r d_a}\left[\alpha_i(L_c - d_a) + \ln\frac{1}{\sqrt{R_{t\alpha}R_{b\alpha}}}\right], \tag{3.13}$$

where the effective length $L_{c,\text{eff}}$ is replaced by the inner cavity length L_c. Knowing the threshold gain, the photon lifetime τ_p is defined as [66]

$$\frac{1}{\tau_p} = v_{\text{gr}}\Gamma\Gamma_r g_{\text{th}} \approx v_{\text{gr}}\left[\alpha_i + \frac{1}{L_{c,\text{eff}}}\ln\frac{1}{\sqrt{R_t R_b}}\right] = v_{\text{gr}}(\alpha_i + \alpha_m), \tag{3.14}$$

with the mirror loss α_m and the approximation assuming the conditions $\alpha_a \ll g_{\text{th}}$, $d_a \ll L_{c,\text{eff}}$ and $\Gamma = \Gamma_z = d_a/L_{c,\text{eff}}$. The group velocity of the laser mode is related to the vacuum velocity of light c as $v_{\text{gr}} = c/\langle \bar{n}_{\text{gr}} \rangle$, and $\langle \bar{n}_{\text{gr}} \rangle$ is the spatial average of the group index $\bar{n}_{\text{gr}} = \bar{n} - \lambda d\bar{n}/d\lambda$.

Comparing the phases in the lasing condition (3.7) leads to the *resonance condition* of the VCSEL emission wavelength λ which can be written as

$$\lambda = \frac{2\pi\langle \bar{n} \rangle L_{c,\text{eff}}}{m\pi + \phi_t + \phi_b}, \tag{3.15}$$

with a positive integer m referring to the *mode order*.

3.3 Operation Characteristics

Under lasing condition, the optical output power [66]

$$P = \frac{\hbar\omega}{q}\eta_\text{d}(I - I_\text{th}) \qquad (3.16)$$

linearly increases with driving current I, as long as temperature effects play no role. Here, I_th is the threshold current, η_d is the differential quantum efficiency which represents the fraction of injected electrons that produce coherent emission in the external beam, and $\hbar\omega$ and q denote photon energy and electron charge, respectively. Considering carrier overflow over confining barriers as well as lateral leakage currents, the current injection efficiency η_I is included through

$$\eta_\text{d} = \tilde{\eta}_\text{d}\eta_\text{I}\,, \qquad (3.17)$$

where $\tilde{\eta}_\text{d}$ is the photonic quantum efficiency which characterizes the percentage of generated coherent light that is available as emission. Commonly, (3.16) is called the light–current (LI) characteristics of the VCSEL from which the power slope efficiency

$$\text{SE} = \frac{\hbar\omega}{q}\eta_\text{d} \qquad (3.18)$$

can be extracted. The threshold current density [66]

$$J_\text{th} = \frac{qd_\text{a}n_\text{th}}{\eta_\text{I}\tau_\text{sp}} \qquad (3.19)$$

is proportional to the threshold carrier density n_th, where the spontaneous recombination lifetime τ_sp given by

$$\frac{1}{\tau_\text{sp}(n)} = \frac{1}{\tau_\text{sp,r}(n)} + \frac{1}{\tau_\text{sp,n}(n)} = A + Bn + Cn^2 \qquad (3.20)$$

depends on the carrier density, with the coefficients A, B and C quantifying non-radiative surface or interface recombination, radiative bimolecular recombination, and Auger recombination, respectively. The radiative lifetime $\tau_\text{sp,r}$ is thus described by B while the non-radiative lifetime $\tau_\text{sp,n}$ is described by A and C. For QWs, the peak gain coefficient can be approximated as

$$g_\text{p}(n) \approx \bar{g}\ln(n/n_\text{t})\,, \qquad (3.21)$$

with the gain constant $\bar{g}$, the active region carrier density n, and the transparency carrier density n_t, at which $g_\text{p}(n = n_\text{t}) = 0$. Generally for VCSELs, there is an offset $\delta\lambda_\text{g}$ between lasing wavelength λ and gain peak wavelength λ_p. To obtain n_th for a given g_th, the shape

3 Fundamentals of VCSELs

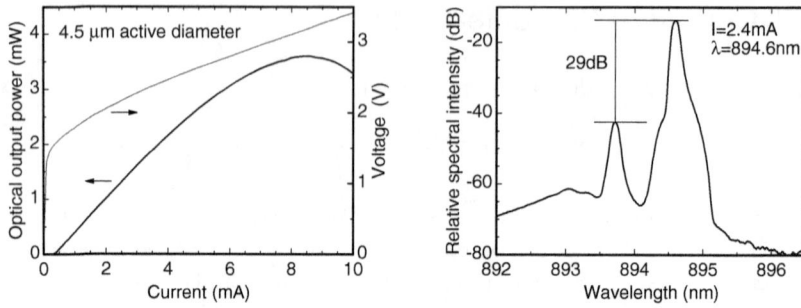

Figure 3.3: Operation characteristics of an 894.6 nm wavelength VCSEL with 4.5 μm active diameter (left). Spectrum of the same VCSEL at 2.4 mA current and $T = 20°C$ ambient temperature.

of the gain spectrum has to be considered. In a simple approach the modal material gain coefficient of the QWs at the spectral position of the lasing mode can be approximated by [66]

$$g(n, \lambda) = g_\mathrm{p}(n)(1 + a_{\mathrm{g}\pm}(\delta\lambda_\mathrm{g})^2), \quad (3.22)$$

with $a_{\mathrm{g}+}$ and $a_{\mathrm{g}-}$ being the curvatures of $g(\lambda)$ at both sides of the gain peak. Assuming perfect alignment with $\delta\lambda_\mathrm{g} = 0$, we have $g_\mathrm{th} = g_\mathrm{p}(n = n_\mathrm{th})$ and the threshold carrier density

$$n_\mathrm{th} = n_\mathrm{t} \exp\{g_\mathrm{th}/\bar{g}\} \quad (3.23)$$

can be determined, where g_th can be obtained from (3.13). If J_th is uniform over the active area A_a, the threshold current [66]

$$I_\mathrm{th} = A_\mathrm{a} J_\mathrm{th} \approx \frac{qV_\mathrm{a}B}{\eta_\mathrm{I}} n_\mathrm{t}^2 \exp\{2g_\mathrm{th}/\bar{g}\} \quad (3.24)$$

shows an exponential growth with an exponent of $2g_\mathrm{th}/\bar{g}$, where the active volume $V_\mathrm{a} = A_\mathrm{a} d_\mathrm{a}$. For simplicity the approximation assumes $A = C = 0$ and the validity of (3.21) and (3.23).

Considering the current–voltage characteristics, the voltage across the VCSEL can be approximated as

$$V \approx V_\mathrm{k} + R_\mathrm{s} I, \quad (3.25)$$

where V_k is the kink voltage and can be approximated by $V_\mathrm{k} \approx \hbar\omega/q$ in optimized VCSELs, and $R_\mathrm{s} = dV/dI$ denotes the differential series resistance. Figure 3.3 (left) shows light–current–voltage (LIV) characteristics of a VCSEL emitting at $\lambda = 894.6$ nm. The threshold current is lower than 0.5 mA and the maximum output power exceeds 3.5 mW at an

3.4 Temperature Behavior

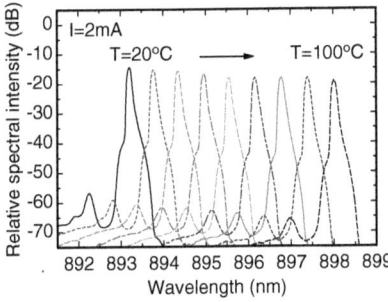
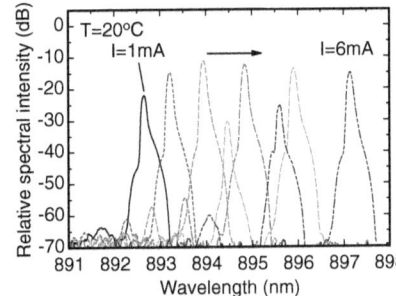

Figure 3.4: Emission spectra of a VCSEL with 2.8 µm active diameter at different substrate temperatures varied between 20 and 100°C in steps of 10°C at a current of 2 mA (left) and for different currents between 1 and 6 mA in steps of 1 mA at a substrate temperature of 20°C (right).

operating current of 8.5 mA. The VCSEL has an SE of 0.6 W/A and an η_d of 43%. Considering the IV curve, one gets $V_k = 1.65$ V and $R_s = 191\,\Omega$. The optical CW spectrum at a current of 2.4 mA is illustrated in Fig. 3.3 (right). The fundamental transverse mode is lasing at 894.6 nm. A higher-order mode is located on the short-wavelength side with a side-mode suppression ratio (SMSR) of more than 25 dB.

3.4 Temperature Behavior

3.4.1 Red-Shift Effect

Because of the short cavity length, the emission wavelength of the VCSEL is determined by the cavity resonance and not by the gain peak as in conventional EELs. Therefore, the thermal shift of the resonance wavelength λ (see (3.15)) is mainly ruled by the changes of the average refractive index in the cavity and subordinately by the thermal expansion of the semiconductor layers. Hence, the thermal wavelength shift is dependent on the material composition of the DBRs and the inner cavity. An increase of the ambient temperature or of the drive current leads to a longer emission wavelength of a VCSEL, as illustrated by the emission spectra plotted in Fig. 3.4 (left) and (right), respectively. This effect is often called *red-shift*. For the VCSEL of Fig. 3.4, the tuning coefficient of the emission wavelength with the substrate temperature is 0.06 nm/K and with the current is 0.9 nm/mA. The red-shift of the emission spectra with increasing current is predominantly dependent on the size of the oxide aperture and implies internal heating of the VCSEL,

3 Fundamentals of VCSELs

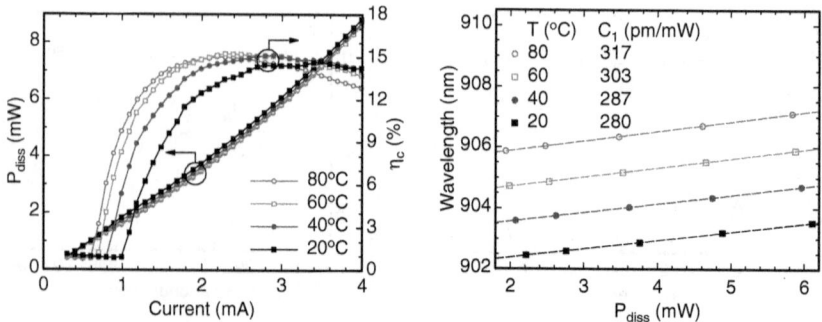

Figure 3.5: Dissipated power P_diss and conversion efficiency η_c in dependence of current for a VCSEL with 3.2 μm active diameter (left). Emission wavelength versus P_diss of the VCSEL along with linear fits at $T = 20$, 40, 60 and 80°C (right).

which is suitably described by the *thermal resistance* in Sect. 3.4.2.

3.4.2 Thermal Resistance

As a practical measure of the VCSEL internal heating, the thermal resistance [66]

$$R_\text{th} = \frac{\Delta T}{P_\text{diss}} = \frac{C_1}{C_2}, \tag{3.26}$$

describes the ratio of the temperature increase $\Delta T = T_\text{int} - T$ in the device and the dissipated electrical power P_diss, where T_int and T are the average temperature in the inner cavity and the substrate temperature[1], respectively. The thermal resistance can be determined from two measurements, in particular the wavelength shift with dissipated power at a constant substrate temperature, $C_1 = \Delta\lambda/\Delta P_\text{diss}|_{T=\text{cons.}}$, and the wavelength shift with varying substrate temperature $C_2 = \Delta\lambda/\Delta T$, usually at pulsed operation to avoid self-heating, i.e., at negligible dissipated power. C_2 is material-dependent and originates from the changes of the average refractive index with temperature $d\bar{n}/dT$. Considering the photon cooling by emitted optical power P, the dissipated electrical power

$$P_\text{diss} = IV - P = IV(1 - \eta_c) \tag{3.27}$$

---

[1] To characterize VCSELs, the wafer is normally mounted on a heat sink, which determines the substrate temperature. A temperature sensor is fixed on the surface of the heat sink as close as possible to the wafer. The reading given by a temperature meter is T, which is not exactly the substrate temperature. Moreover, the temperature is not uniform within the substrate which has nearly 200 μm thickness after thinning. In particular, the temperature close to the n-type DBR increases due to the self-heating in the inner cavity. For simplicity, throughout the dissertation, the measured temperature T is called the substrate temperature.

3.4 Temperature Behavior

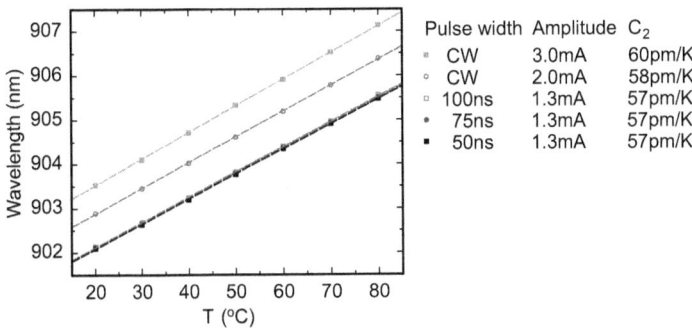

Figure 3.6: Emission wavelength versus T of the VCSEL from Fig. 3.5 measured under CW operation at $I = 2$ and $3\,\text{mA}$ and under pulsed-mode operation with constant pulse amplitude and repetition period of $1.3\,\text{mA}$ and $50\,\upmu\text{s}$, respectively, and varied pulse widths of 100, 75, and $50\,\text{ns}$. The dashed lines are linear fits from which the C_2 factor is extracted.

is current dependent, where η_c is the wallplug or conversion efficiency. The current dependence of P_diss and η_c for a VCSEL with $3.2\,\upmu\text{m}$ active diameter for T varied between 20 and $80°\text{C}$ in steps of $20°\text{C}$ is depicted in Fig. 3.5 (left). Maximum values $\hat{\eta}_c = 14.7\%$, 15.1%, 15.2%, and 15.0% are reached at $I = 3.5, 2.9, 2.3, 2.2\,\text{mA}$ for T varied from 20 to $80°\text{C}$ in steps of $20°\text{C}$, respectively. In a similar way of the measurements shown in Fig. 3.4 (right), the emission wavelengths at different currents of the VCSEL of Fig. 3.5 (left) were measured at several substrate temperatures. The C_1 factor is then obtained as the slope of the linear fits of the emission wavelength plotted against P_diss for different T, as shown in Fig. 3.5 (right). It has values of 280, 287, 303 and $317\,\text{pm/mW}$ for $T = 20$, 40, 60, $80°\text{C}$, respectively.

Like the measurements shown in Fig. 3.4 (left), the emission wavelengths at different T of the VCSEL from Fig. 3.5 (left) were measured under CW operation at $I = 2$ and $3\,\text{mA}$ as well as under pulsed-mode operation[2] using a pulsed-current generator[3], with pulse amplitude of $1.3\,\text{mA}$, pulse period of $50\,\upmu\text{s}$ and different pulse widths of 100, 75, and $50\,\text{ns}$. The pulse signal was displayed using an oscilloscope[4] to determine its amplitude

---

[2] For the pulsed-mode operation, it is necessary to bias the VCSEL with low current (e.g., $I = 20\,\upmu\text{A}$) using a bias-T to avoid the operation at the high-series-resistance region at $I \approx 0$. Otherwise, the input pulse is substantially distorted by a reflected pulse almost having the same amplitude and a phase shift of π due to the large impedance mismatch between the VCSEL and the $50\,\Omega$ measurement system.

[3] Avtech Electrosystems Ltd., model AVPP-2-C-P.

[4] Tektronix, model TDS 3032.

3 Fundamentals of VCSELs

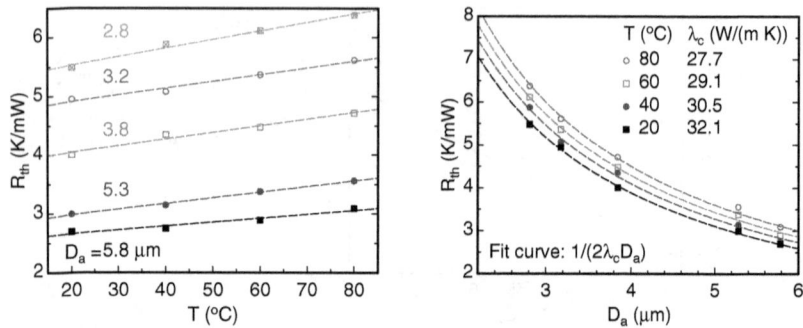

Figure 3.7: Experimental and fitted thermal resistances of VCSELs with different active diameters as a function of substrate temperature T (left) and as a function of active diameter D_a (right).

and width. The C_2 factor is then obtained as the slope of the emission wavelength against T, as shown in Fig. 3.6. For different pulse widths, the C_2 factor is constant and has a value of 57 pm/K. Under CW operation, C_2 increases slightly due to self-heating and becomes 58 and 60 pm/K for $I = 2$ and 3 mA, respectively. As mentioned above, C_2 originates from the variation of the refractive index with temperature $d\bar{n}/dT$. It is thus independent of the active diameter. For the layer structure of all the VCSELs in this work, C_2 is typically found to be around 0.057 nm/K. R_th is then calculated using (3.26).

By measuring the C_1 factors at different substrate temperatures for other VCSELs having different active diameters, the dependence of R_th on the substrate temperature and the device active diameter is obtained, as illustrated in Fig. 3.7. Figure 3.7 (left) depicts R_th versus T for VCSELs having active diameters of 2.8, 3.2, 3.8, 5.3, and 5.8 µm. As can be seen, R_th increases with increasing T and decreased active diameter. This behavior of the thermal resistance follows the simple analytical estimation [74]

$$R_\mathrm{th} \approx \frac{1}{2\lambda_c D_\mathrm{a}}, \tag{3.28}$$

which is obtained assuming a heat flow from a circular area with a diameter D_a, identified here as the active diameter, into a half-space filled with a homogenous medium of a thermal conductivity λ_c. Figure 3.7 (right) depicts thermal resistances of the VCSELs from Fig. 3.7 (left) against their active diameter along with curve fits using (3.28). For instance, a VCSEL with 4 µm active diameter would show a thermal resistance of between 3.9 and 4.5 K/mW for T varied between 20 and 80°C. Extracted values of λ_c are 32.1, 30.5, 29.1, and 27.7 W/(m·K) for $T = 20$, 40, 60, and 80°C, respectively, which are in good agreement with a modeled thermal conductivity of AlGaAs alloys [75]. From the absolute room temperature $T_0 = 293\,\text{K}$ and its associated thermal conductivity $\lambda_{c,0}$, the temperature dependence of λ_c can be estimated by the empirical relation [76]

$$\lambda_c(T) = \lambda_{c,0} \left(\frac{T_0}{T}\right)^{\hat{n}}. \tag{3.29}$$

The exponent $\hat{n} > 0$ is a real number and is found to be approximately 0.78 for $\lambda_c(T)$ extracted from Fig. 3.7 (right) with a maximum error less than 0.25%. The decrease of λ_c explains the increase of C_1 in Fig. 3.5 (right) and consequently the increase of R_{th} in Fig. 3.7 (left) or (right) with increasing T.

3.5 Dynamic and Noise Behavior

The dynamic and noise characteristics of VCSELs are generally obtained from rate equations which quantify the time-dependent change of electrons and holes in the laser cavity as well as their mutual interaction. The small-signal current modulation response is usually used to describe the dynamic behavior. Intrinsically laser noise originates from spontaneous emission. The amplitude noise is evaluated by the RIN spectrum, while the phase noise can be seen as a finite emission linewidth.

3.5.1 Rate Equations

Assuming single transverse mode laser oscillation, the widely accepted laser rate equations in terms of time derivatives $\mathrm{d}/\mathrm{d}t$ are written as [66]

$$\frac{\mathrm{d}n}{\mathrm{d}t} = \frac{\eta_{\mathrm{i}} j}{q d_{\mathrm{a}}} - \frac{n}{\tau_{\mathrm{sp}}(n)} - \Gamma_r g(n) N v_{\mathrm{gr}}, \tag{3.30}$$

$$\frac{\mathrm{d}N}{\mathrm{d}t} = \Gamma \beta_{\mathrm{sp}} \frac{n}{\tau_{\mathrm{sp,r}}(n)} + \Gamma \Gamma_r g(n) N v_{\mathrm{gr}} - \frac{N}{\tau_{\mathrm{p}}}, \tag{3.31}$$

where n is the carrier density and N is the photon density of the mode. The photon density is increased by spontaneous and stimulated emissions according to the spontaneous recombination rate $\Gamma \beta_{\mathrm{sp}} n / \tau_{\mathrm{sp,r}}$ and the stimulated recombination rate $\Gamma \Gamma_r g(n) N v_{\mathrm{gr}}$, respectively, where β_{sp} is the spontaneous emission factor, $\tau_{\mathrm{sp,r}}$ is the radiative part of the spontaneous recombination lifetime given by (3.20), and Γ_r introduced in Sect. 3.1 accounts for the spatial overlap between active region (i.e., QWs) and the standing-wave pattern. The photon density decays due to optical losses and outcoupling which are both included in τ_p according to (3.14).

The modal material gain coefficient of the QWs at the spectral position of the lasing mode can be approximated by [66]

$$g(n, N) = \frac{g(n)}{1 + \bar{\varepsilon}N}, \qquad (3.32)$$

where the parameter $\bar{\varepsilon}$ accounts for gain compression [77] caused mainly by carrier transport and capture effects [78] and to lesser extent by carrier heating and spectral hole burning. $g(n)$ was already introduced in (3.22) and is identical to $g_p(n)$ from (3.21), if the offset $\delta\lambda_g = 0$.

3.5.2 Small-Signal Modulation Response

The small-signal modulation response of VCSELs can be approximated by the three-pole transfer function [66, 79]

$$|M(f)|^2 = \frac{\bar{A}f_r^4}{(f_r^2 - f^2)^2 + (\gamma f/(2\pi))^2} \cdot \frac{1}{1 + (f/f_p)^2}, \qquad (3.33)$$

where $|M|^2$ describes the square of the fluctuation of the optical output power relative to the modulating electrical power and f is the modulation frequency. The first term in (3.33) can be derived by a small-signal analysis of the laser rate equations (3.30) and (3.31) representing the intrinsic carrier–photon interaction, which results in the ideal damping-limited modulation behavior of the semiconductor laser, where $\bar{A}$ is a constant, f_r is the resonance frequency, and γ is the damping coefficient. Neglecting thermal effects and spontaneous emission, the resonance frequency

$$f_r \approx \frac{1}{2\pi}\sqrt{\frac{\eta_i v_{gr} \Gamma_r \bar{a}}{qV_p}} \cdot \sqrt{I - I_{th}} = D \cdot \sqrt{I - I_{th}} \qquad (3.34)$$

is current-dependent and proportional to the so-called D-factor, where v_{gr} is the photon group velocity and $\bar{a}$ is introduced by linearizing the carrier-dependent part of the gain coefficient in (3.21)[5] about an operating point $(n = n_0)$ as

$$g(n_0) = \bar{g}\ln\frac{n_0}{n_t} \approx \bar{a}(n_0 - \bar{n}_t), \qquad (3.35)$$

with the differential gain coefficient $\bar{a}$ and the linearization transparency carrier density $\bar{n}_t$ given by

$$\bar{a} = \frac{\bar{g}}{n_0} \quad \text{and} \quad \bar{n}_t = n_0\left(1 - \ln\frac{n_0}{n_t}\right). \qquad (3.36)$$

---
[5]For simplicity $\delta\lambda_g = 0$ is assumed here.

3.5 Dynamic and Noise Behavior

The frequency at which the response $|M(f)|^2$ has its maximum,

$$f_{\text{peak}} = f_{\text{r}} \sqrt{1 - \frac{1}{2}\left(\frac{\gamma}{2\pi f_{\text{r}}}\right)^2}, \tag{3.37}$$

is slightly less than f_{r} depending on the damping coefficient

$$\gamma = K f_{\text{r}}^2 + \frac{1}{\tau_{\text{sp}}} = K f_{\text{r}}^2 + \gamma_0, \tag{3.38}$$

which is proportional to f_{r}^2 via a proportionality constant known as the K-factor

$$K = 4\pi^2 \left(\tau_{\text{p}} + \frac{\bar{\varepsilon}}{v_{\text{gr}}\Gamma_{\text{r}}\bar{a}}\right), \tag{3.39}$$

where γ_0 is the damping coefficient offset. According to (3.34) and (3.38) f_{r} and γ increase with higher bias current until eventually γ becomes large enough that $|M(f)|^2$ drops by 3 dB at a frequency less than f_{r}. As a figure of merit, the modulation current efficiency factor (MCEF) specifies the increase of the 3 dB corner frequency of the response $|M(f)|^2$ with the bias current as

$$\text{MCEF} = \frac{f_{3\,\text{dB}}}{\sqrt{I - I_{\text{th}}}}. \tag{3.40}$$

In the absence of any parasitic effects, $f_{3\,\text{dB}}$ increases steadily until reaching the theoretical damping-limited maximum 3 dB corner frequency

$$f_{\text{max,d}} = \frac{\sqrt{8}\pi}{K}. \tag{3.41}$$

In order to achieve a large $f_{\text{max,d}}$, the K-factor given by (3.39) shall be decreased either by increasing the differential gain or decreasing the photon lifetime.

Deviations from the intrinsic modulation behavior especially occur due to electrical parasitic elements found in equivalent-circuit model of the laser. The equivalent-circuit model contains the distributed resistances, inductances, and capacitances originating from the structure and geometry of the VCSEL. The second term in (3.33) accounts for the effect of the electrical parasitic elements on the modulation behavior by a simple first-order low-pass filter with 3 dB corner frequency f_{p}. The approximation (3.33) does not account for the laser bandwidth limitation due to carrier transport and capture effects, which are not speed-limiting for the frequency range up to 15 GHz [80]. Therefore, they are ignored in the analysis of the modulation behavior of VCSELs discussed in this dissertation. Nevertheless, they would appear as another first-order low-pass in (3.33) [66].

Besides the electrical parasitic elements, there are other parasitic factors such as inferior thermal management and multi-mode oscillation which constrain the VCSEL from

3 Fundamentals of VCSELs

reaching its $f_{\text{max,d}}$ [81]. For the latter, spreading the optical power equally across $\bar{N}_{\text{mod}}$ transverse modes causes MCEF to be proportional to $1/\sqrt{\bar{N}_{\text{mod}}}$ and as a result the $f_{3\,\text{dB}}$ versus $\sqrt{I - I_{\text{th}}}$ saturates once the VCSEL begins to operate in multiple modes [81]. Higher modulation bandwidths were obtained from single-mode devices for an extended current range above threshold [82,83]. However, a given small oxide aperture, conventionally required for single-mode operation, results in a higher thermal resistance. This leads to more self-heating as the VCSEL is operated at higher I required to reach higher $f_{3\,\text{dB}}$. This thermal effect is found to deviate $f_{3\,\text{dB}}$ and f_{r} versus $\sqrt{I - I_{\text{th}}}$ from their linearity (i.e., MCEF and D begin to decrease) until reaching the maximum frequencies $f_{3\,\text{dB,max}}$ and $f_{\text{r,max}}$ [83]. Both can be experimentally extracted, as will be shown in Chap. 5. The deviation from linearity is due to a thermally-induced reduction of the differential gain coefficient[6] and the current injection efficiency[7] [84, 85]. Assuming no damping and no electrical parasitic effects, the thermally-limited maximum 3 dB corner frequency [86, 87]

$$f_{\text{max,t}} = \sqrt{1 + \sqrt{2}} f_{\text{r,max}} \approx 1.55 f_{\text{r,max}} \qquad (3.42)$$

is calculated from the maximum resonance frequency $f_{\text{r,max}}$.

As a summary of the above discussion about laser modulation behavior, one has to be aware that VCSELs exhibit damping-, electrical parasitic- and thermally-limited maximum corner frequencies quantified by $f_{\text{max,d}}$, f_{p} and $f_{\text{max,t}}$, respectively. Each of the three bandwidth limits gives the true maximum 3 dB corner frequency only if the other two limits are significantly higher. The parasitic effect of multi-mode oscillation does not play a role for VCSELs presented in this dissertation since they are single-mode as required for atomic clock applications.

3.5.3 Intensity Modulation and Frequency Modulation

The modulation of the bias current of a laser diode is associated with variations of the charge carrier density, which fluctuates the refractive index. From the resonance condition of a VCSEL expressed by (3.15), the refractive index fluctuations cause a modulation of the laser emission wavelength or frequency. Of course, the direct effect of current modulation is the modulation of optical output power or intensity. This is related to the modulation of carrier density and thus gain. In total, one can say that intensity modulation (IM)

---

[6]Depending on the initial spectral position of the gain peak relative to the cavity resonance (i.e., $\delta\lambda_{\text{g}}$), $\bar{g}$ and consequently $\bar{a}$ from (3.36) will either decrease or increase with temperature as the gain peak and the cavity resonance become less or more aligned, respectively. Secondly, both $\bar{g}$ and $\bar{a}$ decrease with temperature.

[7]Namely due to thermally-induced leakage of charge carriers out of the QWs.

3.5 Dynamic and Noise Behavior

is always related to frequency modulation (FM) of a laser diode. Such IM–FM coupling causes a distribution of the RF modulation power over several equally separated sidebands (see Fig. 2.1 (c)) and is well quantified by the so-called *Henry factor* (also called *linewidth enhancement factor*, see Sect. 3.5.5) [88]. α_H relates the fluctuations of the refractive index to those of the material gain, where both occurs due to variations of the carrier density in the active region, as [89]

$$\alpha_H = \frac{4\pi}{\Gamma \Gamma_r \bar{a} \lambda} \cdot \frac{\partial \bar{n}}{\partial n}, \tag{3.43}$$

and typically shows values between -2 and -7. In (3.43) Γ_r was added for application to VCSELs. The electric field $\bar{E}_m(t)$ of light emission from a harmonically intensity-modulated VCSEL can be expressed as

$$\bar{E}_m(t) = \bar{E}_0[1 + R\sin(2\pi f t + \phi)]\cos(2\pi f_0 t + M\sin(2\pi f t)), \tag{3.44}$$

where $\bar{E}_0$ is a constant amplitude, f_0 is the optical carrier frequency, f is the modulation frequency, $M > 0$ and $R > 0$ are the frequency and intensity modulation indices, respectively, and ϕ is the relative phase between both modulations [90]. The Henry factor can be determined as [91]

$$\alpha_H = -\frac{M}{R}. \tag{3.45}$$

Expanding (3.44) and doing some manipulations yields

$$I_0 \propto J_0^2(M) + R^2 J_1^2(M)\cos^2(\phi), \tag{3.46}$$

and

$$I_{\pm 1} \propto \left[J_1(M) \pm \frac{R}{2}(J_0(M) + J_2(M))\sin(\phi)\right]^2 + \left[\frac{R}{2}(J_0(M) - J_2(M))\cos(\phi)\right]^2 \tag{3.47}$$

where I_0, I_{+1} and I_{-1} are the spectral power intensities of the carrier and the upper and lower first-order sidebands, respectively. $J_{\bar{k}}$ is the Bessel function of order $\bar{k}$. For $\phi = \pi/2$, the upper first-order sideband has larger intensity than the lower sideband. For $\phi = 3\pi/2$, the lower first-order sideband becomes larger. Symmetric first-order sidebands are obtained for $\phi = 0$ or π. The asymmetry of the sidebands can be expressed by the asymmetry factor

$$S = -10\log\left(\frac{I_{+1}}{I_{-1}}\right). \tag{3.48}$$

For CPT excitation in MACs, symmetric modulation sidebands are favorable for better clock performance [92], since employing asymmetric sidebands causes unequal frequency shifts of the two ground states via the dynamic Stark effect. This causes stronger light shifts by fluctuations of the laser output intensity [48]. Additionally, it causes a distorted CPT signal because of unequal pumping of the two CPT atomic transitions. More symmetric modulation sidebands can be achieved via operating the VCSEL at low output power and modulating it at a frequency higher than its resonance frequency [92].

3.5.4 RIN

Even with a noise-free VCSEL current driver, fluctuations will be present in a VCSEL's steady-state output power because of spontaneous emission. This noise characteristic is conveniently described by the relative intensity noise. In the time domain, RIN relates the noise of the optical power density $\delta P(t)$ having units of $W/\sqrt{Hz}$ to the mean power $\langle P \rangle$ as [93]

$$\text{RIN} = \frac{\langle \delta P^2 \rangle}{\langle P \rangle^2}, \quad (3.49)$$

where the angular brackets denote an average over the observation time and the average-square of $\delta P(t)$ is expressed as

$$\langle \delta P^2 \rangle = \frac{1}{2\Delta f} \int_{-\infty}^{+\infty} \langle |\Delta \tilde{P}(f)|^2 \rangle \mathrm{d}f, \quad (3.50)$$

where $\Delta \tilde{P}(f)$ is roughly the Fourier transform of $\delta P(t)$ and the average-square of its magnitude denotes the statistical variance of the noise spectral density of the laser optical power or simply the noise spectral density (NSD). Δf is the noise bandwidth[8] and the factor of 2 in (3.50) arises because of the fact that positive and negative frequencies have to be considered. If the noise is passing a band-pass filter with a center frequency f, a bandwidth Δf and unity transmission in the pass-band and zero in the stop-band, then the average-square value

$$\langle \delta P^2 \rangle = \frac{2\Delta f \langle |\Delta \tilde{P}(f)|^2 \rangle}{2\Delta f} = \langle |\Delta \tilde{P}(f)|^2 \rangle \quad (3.51)$$

is obtained. Using (3.51), the spectral RIN can be expressed as

$$\text{RIN}(f) = \frac{\langle |\Delta \tilde{P}(f)|^2 \rangle}{\langle P \rangle^2} = \frac{\langle |\Delta \tilde{N}(f)|^2 \rangle}{\langle N \rangle^2} = \frac{\langle |\Delta \tilde{I}_{\text{PD,laser}}(f)|^2 \rangle}{\langle I_{\text{PD}} \rangle^2}, \quad (3.52)$$

where the proportionality between the NSD values of output power $\langle |\Delta \tilde{P}(f)|^2 \rangle$, photon density $\langle |\Delta \tilde{N}(f)|^2 \rangle$, and detector photocurrent $\langle |\Delta \tilde{I}_{\text{PD,laser}}(f)|^2 \rangle$ and correspondingly for

---

[8]Normally the noise bandwidth is equivalent to the measuring system bandwidth.

the mean values $\langle P \rangle$, $\langle N \rangle$, and $\langle I_\text{PD} \rangle$ have been employed. Due to its parasitics-free property, RIN can be employed not only to determine the noise characteristics of a laser, but also to examine its intrinsic modulation behavior. The RIN spectrum can be fit to extract f_r and γ using [66, 94]

$$\text{RIN}(f) = \frac{\bar{A}_1 f^2 + \bar{A}_0}{(f_\text{r}^2 - f^2)^2 + (\gamma f/(2\pi))^2}, \tag{3.53}$$

where $\bar{A}_1$ and $\bar{A}_0$ are constants. Considering RIN measurement system, the above treatment has not accounted for other noise contributions such as thermal noise and shot noise. The former is mainly due to thermal noise of the employed photodiode and its NSD can be expressed in terms of the noise-equivalent power (NEP, measured in units of $\text{W}/\sqrt{\text{Hz}}$) and the responsivity R_PD of the photodiode as[9]

$$\langle |\Delta \tilde{I}_\text{PD,therm}(f)|^2 \rangle = (R_\text{PD} \cdot \text{NEP})^2 = \left(\frac{\eta_\text{PD} q}{\hbar \omega} \cdot \text{NEP} \right)^2, \tag{3.54}$$

where η_PD is the photodiode quantum efficiency. The shot noise is due to the quantum nature of the light and its NSD is expressed as

$$\langle |\Delta \tilde{I}_\text{PD,shot}(f)|^2 \rangle = 2q \langle I_\text{PD} \rangle. \tag{3.55}$$

Considering thermal and shot noise contributions to the laser RIN measurement, the NSD of the total measured photocurrent $\Delta \tilde{I}_\text{PD}(f)$ is defined by

$$\langle |\Delta \tilde{I}_\text{PD}(f)|^2 \rangle = \langle |\Delta \tilde{I}_\text{PD,laser}(f)|^2 \rangle + \langle |\Delta \tilde{I}_\text{PD,therm}(f)|^2 \rangle + \langle |\Delta \tilde{I}_\text{PD,shot}(f)|^2 \rangle. \tag{3.56}$$

Dividing (3.56) by $\langle I_\text{PD} \rangle^2$ gives

$$\text{RIN}_\text{total} = \frac{\langle |\Delta \tilde{I}_\text{PD}(f)|^2 \rangle}{\langle I_\text{PD} \rangle^2} = \text{RIN} + \text{RIN}_\text{therm} + \text{RIN}_\text{shot}, \tag{3.57}$$

where

$$\text{RIN}_\text{therm} = \left(\frac{R_\text{PD} \cdot \text{NEP}}{\langle I_\text{PD} \rangle} \right)^2 = \left(\frac{\eta_\text{PD} q}{\hbar \omega \langle I_\text{PD} \rangle} \cdot \text{NEP} \right)^2 \tag{3.58}$$

and

$$\text{RIN}_\text{shot} = 2q/\langle I_\text{PD} \rangle \tag{3.59}$$

are the contributions of thermal noise and shot noise to RIN_total.

---
[9]This definition assumes that thermal noise is dominating during the determination of the NEP.

3.5.5 Emission Linewidth

The finite linewidth of a single-mode laser emission is caused by the stochastic phase fluctuations of its electric field originating from spontaneous emission processes [66, 79, 89]. The resulting spectral power density at a center frequency f_0 of the mode can be approximated by a Lorentzian lineshape function [66]

$$|\tilde{E}(f)|^2 = |\tilde{E}(f_0)|^2 \frac{(\Delta f_\mathrm{L}/2)^2}{(\Delta f_\mathrm{L}/2)^2 + (f - f_0)^2} \qquad (3.60)$$

with the full linewidth at half maximum

$$\Delta f_\mathrm{L} = \frac{\beta_\mathrm{sp} \Gamma \langle n \rangle}{4\pi \tau_\mathrm{sp} \langle N \rangle}(1 + \alpha_\mathrm{H}^2) + \Delta f_{\mathrm{L}_0}. \qquad (3.61)$$

According to (3.61), the linewidth increases linearly with inverse output power but has a residual linewidth Δf_{L_0} if $P \propto \langle N \rangle$ is extrapolated to ∞.

There are several techniques for the direct measurement of the laser emission linewidth such as delayed self-heterodyne [95], heterodyne using a narrow-linewidth external cavity diode laser [96], and scanning Fabry–Pérot interferometer [97, 98]. The laser linewidth can be indirectly determined from the spectral power density of the laser frequency noise $\tilde{S}_\mathrm{FN}(f)$ [99]. Laser frequency noise is a measure of the fluctuations of the emission frequency around its average value and has a unit of $\mathrm{Hz}^2/\mathrm{Hz}$. The emission linewidth is strongly connected to the frequency noise and can be estimated as

$$\Delta f_\mathrm{L} = \sqrt{8 A_\mathrm{FN} \ln 2}, \qquad (3.62)$$

where A_FN is the integral of $\tilde{S}_\mathrm{FN}(f)$ over f [99].

3.6 Polarization Properties

Owing to the cylindrical symmetry of the VCSEL resonator and the isotropic gain and reflectivity provided by the QWs and the DBRs, respectively, the polarization orientation of the emitted light of a standard VCSEL is a priori unknown. However, there are sources of anisotropy in the VCSEL structure like the *electro-optic effect* by which the internal static electric fields in the VCSEL induce a birefringence (i.e., difference in refractive index) along the [011] and [0$\bar{1}$1] crystal axes in a laser grown on a [100]-oriented GaAs substrate [100]. This effect establishes the main polarization directions which exhibit a relative difference in their emission frequencies and a net modal gain difference [101]. The

3.6 Polarization Properties

net modal gain difference is commonly called *linear modal dichroism* and the polarization mode with the higher net modal gain lases [101]. However, the electro-optic effect is too weak to stabilize the polarization.

Another anisotropy source present in oxide-confined GaAs VCSELs is the difference of the wet-thermal oxidation rates of the AlAs layer along different crystal axes. This leads to a non-circular or rhombic shaped oxide aperture even in case of a circular mesa. As a result, a non-cylindrical waveguide is formed. Additionally, the stable oxide produced by wet-thermal oxidation of the AlAs layer is amorphous Al_2O_3 which exhibits strain at the oxide–semiconductor interface due to volume shrinkage [102]. The strain leads to a birefringence inside the VCSEL via the *elasto-optic effect* [103, 104]. The birefringence appears as a difference of refractive index along GaAs crystal axes parallel or orthogonal to the strain direction. Depending on the direction and the magnitude of the strain, the main polarization directions rotate and become not aligned to the [011] and [0$\bar{1}$1] crystal axes. Moreover, the linear modal dichroism induced originally by the electro-optic effect is also dependent on the internal strain [103]. Besides the oxide aperture, internal strain in VCSELs can originate from other sources such as i) defects inside the epitaxially grown material and ii) handling steps like soldering or bonding, which are, for instance, necessary for the atomic clock microsystem integration. Due to the birefringence introduced by the electro-optic and elasto-optic effects in VCSELs, the emission frequencies of the two polarization modes differ by up to 80 GHz [45]. As a conclusion, there is no general polarization selection rule that could be identified for a standard oxide-confined VCSEL. Often, a majority of standard VCSELs on a wafer has equal polarization. However, the polarization can easily switch to an orthogonal direction upon changes of the bias current or the operating temperature [11, 105–107]. Polarization switches or an undefined polarization in general increase the RIN of VCSELs [12, 108], which is an important performance parameter in atomic clocks. As mentioned in Sect. 2.3.3, such polarization switching noise can moreover distort the CPT signal of a MAC and can hence degrade its short-term stability [44].

Besides atomic clock applications, polarization stability of the VCSEL emission is of great interest for many other applications including other atomic sensors like magnetometers or gyroscopes [10], gas sensors [109, 110], and optical navigation sensors like computer mice which became a mass market for VCSELs [111, 112]. Such optical sensors require VCSELs with a well-defined and stable polarization [113, 114]. A short review of the research attempts which have been undertaken during the last 20 years to stabilize the polarization of VCSELs will be presented in Sect. 4.4, among which the polarization control approach applied in this dissertation will be discussed in more detail.

3 Fundamentals of VCSELs

3.7 VCSEL Applications

VCSELs emitting at 850 nm wavelength became first commercialized in the late 1990s for multi-mode fiber (MMF) optical networks. In such networks, data transmission is done over short distances (e.g., 100 m to few kilometer range) [115, 116]. This application field led to significant improvements of VCSEL performance in terms of conversion efficiency, modulation bandwidth, and reliability. Meanwhile it turned out that VCSELs are also ideal laser sources for optical navigation devices like computer mice [111, 112], which have driven the development of polarization-stable single-mode VCSELs [113, 114]. Today, these two application areas approximately equally share a production volume of about 100 million units per year [57].

Another VCSEL application is gas sensing and analysis employing the TDLAS (tunable diode laser absorption spectroscopy) method [109, 110]. VCSEL-based gas sensors emerge in fields like medicine [117] as well as industry, especially for monitoring and control of combustion processes [118]. The TDLAS is much sensitive (i.e., exhibits stronger absorption) in the long wavelength region ($\lambda > 1.3\,\mu$m) [119]. GaAs-based VCSELs are mainly used for oxygen sensing at about 763 nm wavelength [120]. Applying the TDLAS technique using the red-shift property of VCSELs offers several advantages over other types of diode lasers in terms of wide wavelength tuning range with single-mode emission, short response time, and low power consumption. MACs are another promising application field for VCSELs due to their low threshold currents, high modulation bandwidths, unique packaging capabilities, and favorable beam characteristics [9, 121]. Similar to the use in computer mice, they must feature strictly polarization-stable single-mode emission. This is the main application topic for VCSELs introduced in this dissertation, as mentioned in Chaps. 1 and 2, and thus more discussions and descriptions of such devices are to be presented in the chapters to follow.

In the early 2000s, laser printing became an important application field for single-mode VCSELs emitting at 780 nm [122–124]. The attractive features of VCSELs such as light emission perpendicular to the wafer surface allowing fabrication in densely packed one- and two-dimensional arrays, low threshold currents, and slowly divergent circular beam enable high ppm (i.e., pages per minute) printing speed, high dpi (i.e., dots per inch) printing resolution as well as low power consumption. Another recent application for VCSELs is optical manipulation including guiding, trapping and sorting of particles having micrometer range of size [125]. In combination with microfluidics, VCSELs offer promising tools for fast, contamination-free and cost-effective analysis and handling of micrometer-sized particles, usually dealt with in biology and medicine. Like laser printing, VCSELs can be fabricated in two dimensional arrays for effective optical manipulation [126].

Chapter 4

Design and Fabrication of VCSELs for Miniaturized Atomic Clocks

This chapter describes and illustrates in detail the design procedure of VCSELs to achieve the target emission wavelength of nearly 894.6 nm required for miniaturized cesium-based atomic clocks. The design is based on quantum-mechanical and wave-optical computations. Techniques to realize VCSELs with single-mode, single-polarization emission essential for atomic clocks are also discussed. For polarization control, a previously developed technique relying on the integration of a semiconducting surface grating in the top Bragg mirror of the VCSEL structure is employed. For optimizing the design parameters of the surface grating, advanced electromagnetic simulations based on a fully vectorial, three-dimensional model have been performed by a collaboration partner and their results are presented and discussed. The atomic clock VCSEL layer structure is described. Finally, the chip design of the atomic clock VCSEL and the technological processing are explained, where a flip-chip-bondable design has been realized for the purpose of integration with the atomic clock microsystem.

4.1 Adjustment of Layer Thicknesses

Starting from an already existing 850 nm VCSEL structure previously fabricated in the Institute of Optoelectronics at Ulm University, there are two main subjects to consider for producing 894.6 nm VCSELs for miniaturized cesium-based atomic clocks. First, the adjustment of the thicknesses of the VCSEL layer structure and second, the design of the active region (i.e., QWs). The latter will be introduced in detail in the next section.

Shifting to a longer wavelength of $\lambda = 894.6$ nm requires adjusting the layer thicknesses. In particular, it requires larger layer thicknesses due to two reasons. First, the longer

4 Design and Fabrication of VCSELs for Miniaturized Atomic Clocks

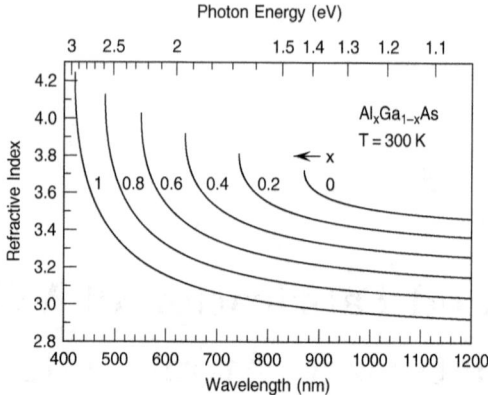

Figure 4.1: Dependence of the refractive index of $Al_xGa_{1-x}As$ on the wavelength (photon energy) for different Al contents x (after the model in Ref. [127]).

wavelength itself and second, the associated reduction in the refractive index due to material dispersion. For instance, the DBRs of the VCSELs in this work are based on the lattice-matched AlGaAs material system. One period of a DBR has a thickness d_B of one half of the emission wavelength in the material, i.e., $d_B = \lambda/(4\bar{n}_H) + \lambda/(4\bar{n}_L)$, with $\bar{n}_H$ and $\bar{n}_L$ as the indices of the high- and low-refractive layers. From Fig. 4.1 one can notice the decrease of the refractive index of $Al_xGa_{1-x}As$ with increasing wavelength. For the same reasons, the one-material-wavelength inner cavity $L_c = \lambda/\bar{n}$ needs to be elongated.

The photon energy hc/λ for $\lambda = 894.6$ nm is less than for $\lambda = 850$ nm. Therefore, the material composition of the DBR can be modified. In particular, the Al content of the high-refractive DBR layer can be reduced, but with keeping its bandgap energy large enough to avoid fundamental absorption of the emitted light. By such a modification, the refractive index step $\Delta\bar{n}_B = \bar{n}_H - \bar{n}_L$ increases. This leads to higher mirror reflectivity compared to a DBR design for 850 nm wavelength, given the same number of the layer pairs. However, in the VCSELs fabricated for this dissertation, the material composition of the DBRs was kept unchanged from the existing 850 nm wavelength design. By this means, a more reproducible epitaxial growth of 850 nm VCSELs was preserved for other users of the epitaxy machine in the Institute.

4.2 Design of the Active Region

The QW structure in the active region of the VCSEL has to be adjusted in a way that the bandgap energy of the QWs fits to the new emission wavelength. For that purpose, indium was introduced into the AlGaAs material system used so far for the QWs of the 850 nm VCSELs. Quantum-mechanical computations using the Schrödinger equation were done in order to evaluate the required indium content. The influence of compressive strain, the band offset of an InGaAs/AlGaAs heterojunction, as well as the effect of bandgap renormalization (which is a many-body effect and depends on the carrier density) have been considered. In the following sub-sections, the approach is described in detail.

4.2.1 Bandgap Energy of Bulk AlGaAs and InGaAs

The VCSELs for miniaturized atomic clock applications are based on the AlGaAs and InGaAs material systems. Therefore one needs to express the bandgap energies of both material systems for different material compositions and ambient temperatures.

AlGaAs

The bandgap energy of $Al_xGa_{1-x}As$ at the Γ-point for 297 K temperature is [128]

$$E_g^\Gamma(x) = (1.424 + 1.247 \cdot x)\,\text{eV} \quad \text{for } 0 \leq x \leq 0.45, \quad (4.1a)$$

$$E_g^\Gamma(x) = (1.424 + 1.247 \cdot x + 1.147 \cdot (x - 0.45)^2)\,\text{eV} \quad \text{for } 0.45 \leq x \leq 1. \quad (4.1b)$$

For Al contents exceeding 0.45, $Al_xGa_{1-x}As$ turns from a direct to an indirect semiconductor. Specifically, the conduction band minimum shifts from the Γ-point to the X-point. The associated bandgap energy at the X-point for 297 K temperature becomes

$$E_g^X(x) = (1.900 + 0.125 \cdot x + 0.143 \cdot x^2)\,\text{eV}\,. \quad (4.2)$$

Our starting approach is to retain the $Al_{0.27}Ga_{0.73}As$ quantum barrier (QB) of 850 nm VCSELs and to calculate the required indium content in the QW in order to tune the optical gain peak to coincide with the wavelength of the cavity resonance at the operating temperature. The expected temperature of the VCSEL in the MAC-TFC demonstrators ranges between 65 and 80°C. Varshni has described the temperature dependence of the bandgap energy by the empirical relation [129]

$$E_g(T) = E_g(0) - \frac{\hat{\alpha}T^2}{\hat{\beta} + T}\,, \quad (4.3)$$

Table 4.1: Material parameters for the computation of the temperature-dependent bandgap energy of GaAs [132] and InAs [133].

Parameter (unit)	GaAs	InAs
$E_g(0)$ (eV)	1.517	0.415
$\hat{\alpha}$ (10^{-4} eV/K)	5.5	2.76
$\hat{\beta}$ (K)	225	83

where T is in Kelvin and $\hat{\alpha}$ and $\hat{\beta}$ are material-dependent constants. For AlGaAs with $x = 0.27$, $E_g(0)$, $\hat{\alpha}$ and $\hat{\beta}$ take values of 1.932 eV, 0.658 meV/K and 248 K, respectively [130].

InGaAs

The bandgap energy of $In_xGa_{1-x}As$ depends on composition and temperature as

$$E_g(x,T) = \left[E_g^{GaAs}(x,0) - \frac{\hat{\alpha}^{GaAs}T^2}{\hat{\beta}^{GaAs} + T} \right] (1-x) +$$

$$\left[E_g^{InAs}(x,0) - \frac{\hat{\alpha}^{InAs}T^2}{\hat{\beta}^{InAs} + T} \right] x - 0.475x(1-x) \, eV, \quad (4.4)$$

which is a quadratic interpolation between GaAs and InAs taking into account the bandgap bowing via the bandgap bowing parameter $E_{g,bow} = -0.475$ eV [131]. $E_g^{InAs}(x,0)$ and $E_g^{GaAs}(x,0)$ are the bandgap energies at 0 K temperature. The values of $E_g(0)$, $\hat{\alpha}$ and $\hat{\beta}$ for GaAs and InAs are depicted in Table 4.1.

4.2.2 Mechanical Strain Effect

Strain

For lattice-mismatched InGaAs QWs grown between AlGaAs QBs, one needs to consider the induced mechanical strain in the bandgap energy calculation. When a layer with a lattice parameter a_L is grown on a substrate with a lattice parameter a_S, either the original lattice parameter of the grown layer remains unchanged, or the lattice parameter of the grown layer adapts to the lattice parameter of the substrate. The first growth is called *incoherent*, while the second one is referred to as *coherent or pseudomorphic growth*. The coherent growth causes the grown layer to be biaxially strained in the plane of the

4.2 Design of the Active Region

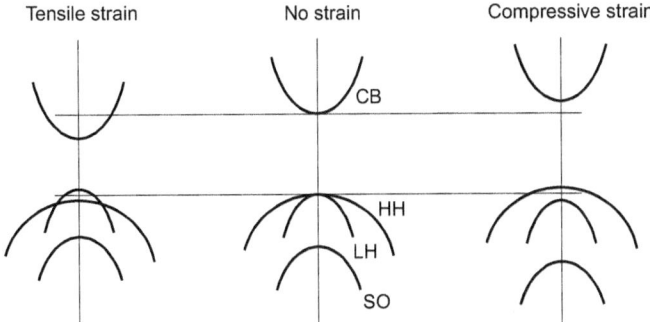

Figure 4.2: Schematic of energy band shifts of the conduction band and the three valence sub-bands by strain. CB: conduction band, LH: light-hole sub-band, HH = heavy-hole sub-band, SO = spin-orbit split-off sub-band (after [135]).

substrate by the strain [134]

$$\varepsilon_\parallel = \frac{a_S - a_L}{a_L} \qquad (4.5)$$

and uniaxially strained in the perpendicular direction by the strain

$$\varepsilon_\perp = -\frac{\varepsilon_\parallel}{\sigma_P}, \qquad (4.6)$$

where the constant $\sigma_P = C_{11}/(2C_{12})$ is known as Poisson's ratio, and C_{11} and C_{12} are the elastic stiffness coefficients. For $a_L > a_S$, the lattice parameter of the grown layer has to shrink, the grown layer is compressively strained, and $\varepsilon_\parallel$ is negative. For $a_L < a_S$, the lattice parameter of the layer must increase, the grown layer has tensile strain, and $\varepsilon_\parallel$ is positive. The introduction of strain lifts the degeneracy at the valence band maximum, i.e., it separates the heavy-hole (HH) and light-hole (LH) sub-bands, as schematically depicted in Fig. 4.2. The lifted degeneracy leads to preferential polarization of the emitted light. For tensile strain, transitions between electrons and light holes lead to TM-polarization. For compressive strain, transitions between electrons and heavy holes lead to TE-polarization. For the pseudomorphic growth, the strain energy grows as the thickness of the grown layer increases. After reaching a critical layer thickness, the strain can no longer be maintained. The lattice mismatch then results in crystalline defects (or dislocations). The critical thickness can be approximated by [134]

$$d_c \approx \frac{a_S}{2|\varepsilon_\parallel|}. \qquad (4.7)$$

4 Design and Fabrication of VCSELs for Miniaturized Atomic Clocks

Table 4.2: Material parameters for the computation of the conduction band shift [137]; a is the lattice parameter.

Parameter (unit)	GaAs	AlAs	InAs	$Al_xGa_{1-x}As$	$In_xGa_{1-x}As$
C_{11} (GPa)	1221	1250	832.9	$1221 + 29x$	$1221 - 388.1x$
C_{12} (GPa)	566	534	452.6	$566 - 32x$	$566 - 113.4x$
$\bar{a}_c$ (eV)	-7.17	-5.64	-5.08	$-7.17 + 1.53x$	$-7.17 - 0.52x + 2.61x^2$
a (Å)	5.6533	5.6611	6.0583	$5.6533 + 0.0078x$	$5.6533 + 0.405x$

For instance, in the present work, the QWs and QBs in some VCSELs are made of $In_{0.06}Ga_{0.94}As$ and $Al_{0.27}Ga_{0.73}As$, respectively. For this combination, $\varepsilon_\| = -0.0039$, and the critical thickness is approximately 72 nm, which is sufficiently large compared to the total QW thickness of 3×8 nm in the present work.

Influence of Strain on the Bandgap Energy

The strain of the QWs affects the optoelectronic characteristics of the semiconductor, which is expressed in a modification of the conduction and valence band edges. In what follows, the influences of strain on the conduction and valence bands are treated separately.

Influence on the conduction band edge:
Under the assumption that the strain is relatively small, i.e., $|\varepsilon_\|| < 1\%$, the first-order perturbation theory can be applied to calculate the band shift [136]. For the case of a cubic crystal structure grown on the [100] direction (as is the case for strained $In_xGa_{1-x}As$ grown on AlGaAs), the shift of the direct conduction band minimum at the Γ-point due to the strain is calculated by [136]

$$\delta E_C = 2\bar{a}_c \left(1 - \frac{C_{12}}{C_{11}}\right) \varepsilon_\|, \qquad (4.8)$$

where $\bar{a}_c$ is the deformation potential constant. It should be mentioned here that the energy shift δE_C is calculated relative to the lower edge of the conduction band at which the energy is zero. For instance, a positive value of δE_C means that the minimum of the conduction band is shifted upwards, like for the case of compressive strain depicted in Fig. 4.2. Table 4.2 lists the needed material parameters.

Influence on the valence band edge:
The computation of the valence band shift due to strain is more complicated, since all three sub-bands (HH, LH, SO) within the valence band should be considered. The computations

4.2 Design of the Active Region

Table 4.3: Material parameters for the computation of the valence sub-band shifts [137].

Parameter (unit)	GaAs	AlAs	InAs	$\text{Al}_x\text{Ga}_{1-x}\text{As}$	$\text{In}_x\text{Ga}_{1-x}\text{As}$
$\bar{a}_\text{v}$ (eV)	-1.16	-2.47	-1.00	$-1.16 - 1.31x$	$-1.16 + 0.16x$
b (eV)	-2.0	-2.3	-1.8	$-2.0 - 0.3x$	$-2.0 + 0.2x$
Δ_SO (eV)	0.341	0.28	0.39	$0.341 - 0.061x$	$0.341 - 0.101x + 0.15x^2$

are based on the so-called $k \cdot p$ perturbation theory [136]. The energies of the band edges of the heavy-hole and light-hole sub-bands at the Γ-point are given by

$$\delta E_\text{HH} = P_\varepsilon + Q_\varepsilon \tag{4.9}$$

and

$$\delta E_\text{LH} = P_\varepsilon - \frac{1}{2}\left(Q_\varepsilon - \Delta_\text{SO} + \sqrt{\Delta_\text{SO}^2 + 2\Delta_\text{SO}Q_\varepsilon + 9Q_\varepsilon^2}\right), \tag{4.10}$$

respectively, where parameters

$$P_\epsilon = -2\bar{a}_\text{v}\epsilon_{\|}\left(1 - \frac{C_{12}}{C_{11}}\right) \tag{4.11}$$

and

$$Q_\epsilon = -b\epsilon_{\|}\left(1 + 2\frac{C_{12}}{C_{11}}\right) \tag{4.12}$$

are introduced. Here, Δ_SO is the spin-orbit splitting, i.e., the energy difference between the HH (or LH) sub-band and the SO sub-band at the Γ-point, and $\bar{a}_\text{v}$ and b are Pikus–Bir deformation potentials describing the influence of the hydrostatic and uniaxial strain, respectively. It should be mentioned here that the energy level shifts δE_HH and δE_LH are calculated downwards starting from the upper edge of the valence band at which the energy is zero. For instance, a positive value of δE_HH means that the heavy-hole sub-band is shifted downwards, as for the tensile strain case depicted in Fig. 4.2. The material parameters are collected in Table 4.3. After calculating the shifts of the conduction and of the valence sub-bands, one can determine the new bandgap energy of the strained material as the difference between the conduction band edge and the topmost valence sub-band edge. The relative changes of the conduction band and the valence sub-bands for a direct semiconductor are represented schematically in Fig. 4.2 for the cases of unstrained, tensile strained and compressively strained layers. In case of tensile strain, the HH sub-band will be lower than the LH sub-band, and vise versa in case of compressive strain. In the present work, $\text{In}_x\text{Ga}_{1-x}\text{As}/\text{Al}_{0.27}\text{Ga}_{0.73}\text{As}$ QWs experience compressive strain for indium contents $x > 0.52\%$. As described above, in case of compressive strain, the HH sub-band is the topmost valence sub-band. The strained bandgap energy is thus

$$E_{\text{g,strained}} = E_{\text{g,unstrained}} + \delta E_{\text{C}} + \delta E_{\text{HH}} \,. \tag{4.13}$$

4.2.3 Bandgap Renormalization

The computations of the bandgap energy so far refer to individual electron–hole transitions. However, for an operating laser, a large number of charge carriers must be injected into the QWs so that stimulated recombination takes place. This many-body effect induces a significant reduction of the bandgap energy which is called *bandgap renormalization* [143]. Assuming that the bandgap renormalization applies equally to all sub-bands, the bandgap energy can be written as

$$E_{\text{g,renorm}} = E_{\text{g}} + \delta E_{\text{g}} \,. \tag{4.14}$$

Neglecting the Coulomb attraction between electrons and holes, the bandgap correction factor

$$\delta E_{\text{g}} = -1.5 \cdot 10^{-8} \sqrt[3]{n/\text{cm}^{-3}} \,\text{eV} \tag{4.15}$$

depends on the charge carrier density n and, as an approximation, is used for all InGaAs QW thicknesses [144]. The VCSELs in this work are found to have $n \approx 10 \cdot 10^{18}\,\text{cm}^{-3}$ resulting in $\delta E_{\text{g}} = -32\,\text{meV}$, which corresponds to a red-shift of the bandgap wavelength of 21 nm at $\lambda = 895\,\text{nm}$[1].

4.2.4 Relative Band Offset

As depicted in Fig. 4.3, the alignment of band edges in adjacent layers of a heterostructure can be described by the energy differences ΔE_{C} and ΔE_{V}, called *relative band offsets*. In order to determine them, one needs a common point of reference. For instance, the band offsets of binary and ternary compound semiconductors are stated with respect to the upper edge of the valence band of InSb [137]. This particular reference point is chosen since InSb possesses the smallest bandgap and highest upper valence band edge compared to other important binary compound semiconductors, as can be noticed in Fig. 4.4.

---

[1]From (3.19) and (3.20), assuming that $A = C = 0$ and $n \approx n_{\text{th}}$ (i.e., in lasing condition), the carrier density can be approximated as $n \approx \sqrt{\eta_{\text{l}} J_{\text{th}}/(q d_{\text{a}} B)}$. Given that $B = 10^{-10}\,\text{cm}^3/\text{s}$, $\eta_{\text{l}} = 0.9$, $d_{\text{a}} = 24\,\text{nm}$ and a practical value of $J_{\text{th}} = 4\,\text{kA/cm}^2$, one obtains $n \approx 10 \cdot 10^{18}\,\text{cm}^{-3}$. However, with this procedure no consideration for carrier-density-dependent effects like carrier overflow and current spreading is taken. Hence, the above value of n can be regarded as an upper limit.

4.2 Design of the Active Region

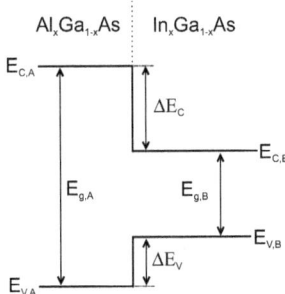

Figure 4.3: Schematic band diagram of an $\text{Al}_x\text{Ga}_{1-x}\text{As}/\text{In}_x\text{Ga}_{1-x}\text{As}$ heterostructure.

Table 4.4: Valence band offsets relative to the level of the upper edge of the valence band of InSb [137].

Parameter (unit)	GaAs	AlAs	InAs	$\text{Al}_x\text{Ga}_{1-x}\text{As}$	$\text{In}_x\text{Ga}_{1-x}\text{As}$
VBO (eV)	−0.80	−1.33	−0.59	$-0.80 - 0.53x$	$-0.8 + 0.59x - 0.38x^2$

For ternary compound semiconductor materials, valence band offsets (VBOs) and conduction band offsets (CBOs) are shown in Fig. 4.5 (left) and (right), respectively. The points indicate offsets for binary and ternary compounds, while the vertical lines signify the band offset ranges that are available using lattice-matched quaternaries. The dashed lines illustrate valence or conduction band offset and lattice constant variations for a number of important ternary alloys.

For the heterostructure of interest for this work, which is made of $\text{In}_x\text{Ga}_{1-x}\text{As}$ and $\text{Al}_x\text{Ga}_{1-x}\text{As}$, Table 4.4 depicts the relative level of the upper valence band edge. As above, the values are with respect to the valence band maximum (VBM) of InSb. The valence band offsets stated in the table and in the figures neglect the influence of the strain in lattice-mismatched heterostructures. The strain influence on the bandgap energy has been considered in Sect. 4.2.2. The uncertainty of the valence band offset in a strained heterostructure originates from the lifted degeneracy of the valence sub-bands. However, for an InGaAs/AlGaAs heterostructure, the influence of band offset uncertainty on the bandgap energy has been found to be subordinate [138]. For $\text{In}_x\text{Ga}_{1-x}\text{As}$, the valence band offset is not linearly dependent on the indium content. For this reason, a bowing parameter $\text{VBO}_{\text{bow}} = -0.38$ eV is involved, as can be seen in Table 4.4. Using the valence band offsets VBO_w and VBO_b of the materials of the QW and the QB, respectively, one

51

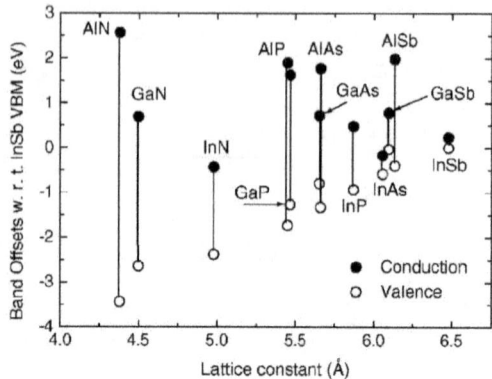

Figure 4.4: Conduction band (bullets) and valence band (circles) offsets at the Γ-point of 12 binary compound semiconductors with respect to the valence band maximum (VBM) of InSb. The extent of the bandgap energy for each material is indicated by a vertical line (Reprinted with permission from [137]. Copyright 2001, AIP Publishing LLC).

can calculate the relative band offset of the valence band

$$\Delta E_{\text{V}} = |\text{VBO}_{\text{b}} - \text{VBO}_{\text{w}}| \tag{4.16}$$

and the relative band offset of the conduction band

$$\Delta E_{\text{C}} = E_{\text{g,b}} - E_{\text{g,w}} - \Delta E_{\text{V}}, \tag{4.17}$$

where $E_{\text{g,w}}$ and $E_{\text{g,b}}$ are the bandgap energies of the QW and QB, respectively. Applying the VBO data from Table 4.4, one obtains $\text{VBO}_{\text{w}} = -0.78\,\text{eV}$ and $\text{VBO}_{\text{b}} = -0.94\,\text{eV}$ for an $\text{In}_{0.04}\text{Ga}_{0.96}\text{As}/\text{Al}_{0.27}\text{Ga}_{0.73}\text{As}$ heterostructure for $T = 293\,\text{K}$. According to (4.16) and (4.17), the relative band offsets $\Delta E_{\text{V}} = 0.166\,\text{eV}$ and $\Delta E_{\text{C}} = 0.323\,\text{eV}$ are found, where $E_{\text{g,b}} = 1.83\,\text{eV}$ is obtained using (4.3) and $E_{\text{g,w}} = 1.34\,\text{eV}$ from (4.4), (4.13) and (4.14) with the effects of compressive strain as well as bandgap renormalization included.

4.2.5 Quantum Effect

Efficient recombination of electron–hole pairs can be achieved by sandwiching a thin layer of a semiconductor material between cladding layers with larger bandgap energy to form a double heterostructure. As the layer thickness in a double heterostructure gets close to the de-Broglie wavelength (e.g., about 10 nm for semiconductor material), quantum effects

4.2 Design of the Active Region

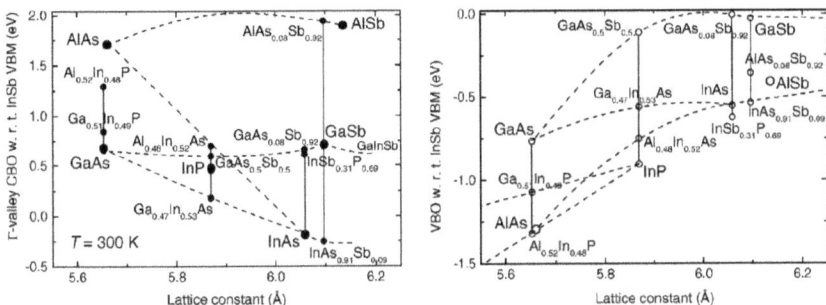

Figure 4.5: Valence band offset (left) and conduction band offset (right) as a function of lattice constant. The offsets for binaries and lattice-matched ternaries are indicated by points. Offset variations with composition for lattice-mismatched ternaries (not including strain effects) are given by dashed lines, and the offset ranges for quaternary alloys lattice-matched to a particular substrate material (GaAs, InP, InAs, or GaSb) are given by the vertical solid lines (Reprinted with permission from [137]. Copyright 2001, AIP Publishing LLC).

become apparent. Mainly, the levels of the allowed energy states become discrete and are not continuous anymore. QWs are important in semiconductor lasers because they allow some degree of freedom in the design of the emission wavelength through adjustment of the energy levels within the QW by careful consideration of the well width and the material composition. Figure 4.6 depicts a simple model of the discrete energy levels in a symmetric QW along with the electron wave functions $\psi(x)$. In order to compute the quantized energy levels, the time-independent Schrödinger equation [136]

$$-\frac{\hbar^2}{2m^*}\frac{\partial^2 \psi}{\partial z^2} + V(z)\psi = E\psi \qquad (4.18)$$

for regions I, II and III in Fig. 4.6 has to be solved. $\hbar = h/(2\pi)$ is the reduced Planck constant and m^* is the effective mass. $V(z)$ is the potential profile of the QW with V_w potential depth (as shown in Fig. 4.6) and E is the energy eigenstate in the QW, where $0 < E < V_\text{w}$. A general solution is [136]

$$\psi_\text{I}(z) = A_1 \exp(\kappa z) + A_2 \exp(-\kappa z) \qquad \text{for} \quad z \leq -\frac{d_\text{w}}{2}, \qquad (4.19\text{a})$$

$$\psi_\text{II}(z) = A_3 \sin(kx) + A_4 \cos(kx) \qquad \text{for} \quad -\frac{d_\text{w}}{2} \leq z \leq \frac{d_\text{w}}{2}, \qquad (4.19\text{b})$$

$$\psi_\text{III}(z) = A_5 \exp(\kappa z) + A_6 \exp(-\kappa z) \qquad \text{for} \quad z \geq \frac{d_\text{w}}{2} \qquad (4.19\text{c})$$

with

$$k = \sqrt{\frac{2m_w^* E}{\hbar^2}} \quad \text{and} \quad \kappa = \sqrt{\frac{2m_b^*(V_w - E)}{\hbar^2}}, \qquad (4.20)$$

where k is the wavenumber, and m_w^* and m_b^* are the effective masses of the carriers in the QW and in the QB, respectively. For confined states, the growing exponential terms in the QB regions must vanish, i.e., $A_2 = A_5 = 0$. As boundary conditions, both $\psi(z)$ and $(1/m^*) \cdot (\partial \psi(z)/\partial z)$ at the interfaces $z = \pm d_w/2$ must be continuous [139, 140], where the latter assures the continuity of current. These two conditions are knows as the Bendaniel–Duke boundary conditions [136]. For the QW region, one can distinguish between two types of solutions. The first is called *even parity*, which is obtained by taking only the cosine function in (4.19b). By applying the continuity conditions, the equation

$$\frac{k}{m_w^*} \tan\left(k \frac{d_w}{2}\right) - \frac{\kappa}{m_b^*} = 0 \qquad (4.21)$$

is obtained. The second solution is called *odd parity*, which is obtained with only the sine function in (4.19b), yielding

$$\frac{k}{m_w^*} \cot\left(k \frac{d_w}{2}\right) + \frac{\kappa}{m_b^*} = 0. \qquad (4.22)$$

Equations (4.21) and (4.22) allow the computation of the discrete energy states for the electrons in the conduction band and for the holes in valence band (with the corresponding effective masses and the appropriate barrier height in (4.20)). One can thus obtain the ground states E_{C1} and E_{V1} of the conduction and the valence band, respectively, from (4.21). The resulting effective bandgap energy is then

$$E_{g,QW} = E_g + E_{C1} + E_{V1}. \qquad (4.23)$$

The associated bandgap wavelength is given by

$$\lambda_g = \frac{hc}{E_{g,QW}} \qquad (4.24)$$

and is indicated in Fig. 4.7, which shows a typical optical gain spectrum of a QW together with a cavity resonance wavelength λ of the fundamental mode of a VCSEL.

The design target of the QW is to achieve the best match at elevated ambient temperatures (65 to 80°C) between the resonance wavelength λ and the gain peak at λ_p. This results in minimum laser threshold currents, as will be illustrated in Sect. 4.2.6. Thus a precise QW design requires the calculation of λ_p rather than λ_g. Basically λ_p is shorter than λ_g by some nanometers. Determining λ_p from λ_g is not a straightforward task and depends

4.2 Design of the Active Region

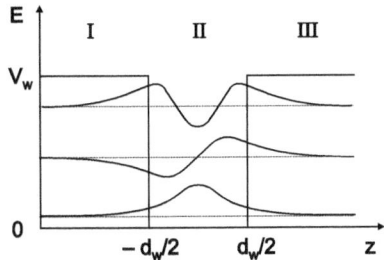

Figure 4.6: Schematic of a one-dimensional symmetric QW with a finite potential V_w and a spatial width d_w. The wave functions for the first three quantized energy levels are depicted. The zero-levels of the wave functions are shifted to correspond to the energy levels in the well.

on different parameters including temperature, charge carrier density and QW thickness [144]. Calculations of λ_p require the computation of the optical gain spectrum, which is beyond the scope of this work. The calculations done so far to determine λ_g are quite useful and can still be utilized to select initial design parameters of the QWs incorporated in the VCSEL structure in order to achieve the desired emission wavelength of 894.6 nm. An increase of the ambient temperature or the drive current lead to a red-shift of the emission wavelength of a VCSEL. Matching the cavity resonance with the gain peak requires the knowledge of the thermal tuning coefficients of their wavelengths. The tuning coefficient of the resonance wavelength $\Delta\lambda/\Delta T > 0$ is due to the positive refractive index change with temperature, $\Delta\bar{n}/\Delta T > 0$. On the other hand, a bandgap shrinkage according to $\Delta E_g/\Delta T < 0$ which corresponds to $\Delta\lambda_g/\Delta T > 0$ induces a red-shift of the overall gain spectrum as indicated in Fig. 4.7. Practical thermal tuning coefficients are $\Delta\lambda/\Delta T \approx 0.06$ nm/K (see Fig. 3.6) and $\Delta\lambda_g/\Delta T \approx 0.3$ nm/K (as can be also calculated from (4.4)) for GaAs-based VCSELs. With increasing temperature, the gain spectrum shifts approximately five times faster to longer wavelengths than the cavity resonance. Therefore the QW should be engineered in a way to locate the cavity resonance on the long-wavelength side of the gain spectrum at room temperature. By this means, matching the cavity resonance with the gain peak can be obtained at elevated ambient temperature and hence a minimum threshold current is achieved. As mentioned earlier in Sect. 2.3.4, the operating temperature for atomic clock VCSELs is ranging between 65 and 80°C. As an example, for a VCSEL to achieve a minimum threshold current at 80°C, it has to emit at $\lambda = 891$ nm at room temperature (e.g., 20°C), while the gain peak should be at $\lambda_p = 876.6$ nm. By elevating the laser temperature up to 80°C, λ and λ_p redshift by $60\,\text{K} \cdot 0.06\,\text{nm/K} = 3.6$ nm and $60\,\text{K} \cdot 0.3\,\text{nm/K} = 18$ nm, respectively, until both

55

4 Design and Fabrication of VCSELs for Miniaturized Atomic Clocks

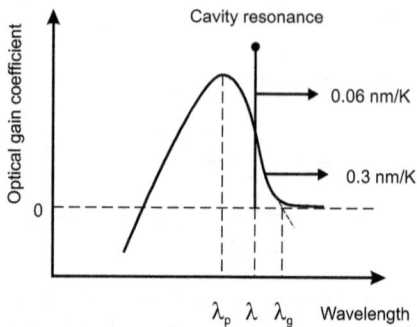

Figure 4.7: Typical optical gain spectrum and cavity resonance together with their thermal tuning coefficients for a GaAs-based VCSEL. λ_p, λ and λ_g are the gain peak, cavity resonance and bandgap wavelengths, respectively.

wavelengths match, yielding $\lambda = \lambda_p = 894.6\,\text{nm}$ at 80°C.

Effective Masses

For the above calculations, the effective masses of electrons and holes at the Γ-point are needed. For unstrained $\text{Al}_x\text{Ga}_{1-x}\text{As}$, the effective masses

$$m_e^* = 0.067 + 0.083 \cdot x, \tag{4.25a}$$

$$m_{\text{HH}}^* = 0.62 + 0.14 \cdot x, \tag{4.25b}$$

$$m_{\text{LH}}^* = 0.087 + 0.063 \cdot x \tag{4.25c}$$

are determined through linear interpolation between the effective masses of GaAs and AlAs [127]. In a similar way, one can calculate the effective masses [141]

$$m_e^* = 0.067 - 0.044 \cdot x, \tag{4.26a}$$

$$m_{\text{HH}}^* = 0.62 - 0.02 \cdot x, \tag{4.26b}$$

$$m_{\text{LH}}^* = 0.087 - 0.06 \cdot x \tag{4.26c}$$

of unstrained $\text{In}_x\text{Ga}_{1-x}\text{As}$. Since in the present work the QWs are compressively strained, the degeneracy of the valence sub-bands is lifted and the HH valence sub-band will have the topmost energy level compared to the LH and SO valence sub-bands. Therefore, we have to use the effective heavy-hole mass in the solution of the Schrödinger equation. However, the effective mass of heavy holes becomes a little bit smaller under compressive

4.2 Design of the Active Region

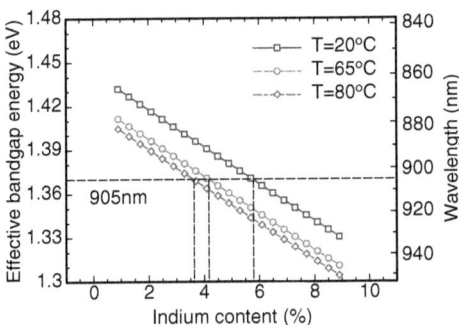

Figure 4.8: Calculated effective bandgap energy and wavelength of an 8 nm thick $In_xGa_{1-x}As/Al_{0.27}Ga_{0.73}As$ QW in dependence of the indium content x for 20, 65 and 80°C ambient temperatures.

strain compared to the unstrained case [142]. According to

$$m^* = \hbar^2 \cdot \left(\frac{d^2E}{dk^2}\right)^{-1}, \qquad (4.27)$$

this means that the curvature of the HH valence sub-band becomes stronger in the vicinity of the Γ-point. Computing the effective mass of the holes in the presence of strain is beyond the scope of this work. For simplicity we apply (4.26a) and (4.26b) also for strained InGaAs QWs.

4.2.6 Experimental Verification

The effective bandgap energy from (4.23) and the associated bandgap wavelength from (4.24) in dependence of the indium content x of an $In_xGa_{1-x}As/Al_{0.27}Ga_{0.73}As$ QW are plotted in Fig. 4.8 for 20, 65 and 80°C ambient temperatures. The influence of compressive strain, the band offsets of the heterojunction, as well as the effect of bandgap renormalization are included. In order to achieve $\lambda_p = 894.6$ nm, λ_g has to be slightly longer and has been assumed to be 905 nm, as indicated by a horizontal dashed line in Fig. 4.8. It can be noticed that $\lambda_g = 905$ nm is reached with approximately $x = 5.8\%$ at 20°C. Since the bandgap energy decreases with increasing temperature, the same bandgap wavelength of 905 nm requires approximately $x = 4.2\%$ and 3.6% indium content at 65 and 80°C, respectively. As an experimental verification of the quantum-mechanical computation results illustrated by Fig. 4.8, three atomic clock VCSEL wafers with $x = 6\%$, 4.5% and 4% have been fabricated. The threshold currents of three VCSELs with different x but

57

identical resonance wavelengths at $T = 20°C$ are illustrated in Fig. 4.9 (left) as a function of T. For constant x, the change of threshold current with temperature can be attributed mainly to the change of the material gain itself and the change of detuning between cavity resonance and gain peak, whereas the mirror reflectivities remain almost constant. A higher x reduces the bandgap energy and thus shifts the optical gain spectrum to longer wavelengths. Consequently, x changes the alignment between the cavity resonance (or emission wavelength) and the gain peak and hence the temperature at which the threshold current is minimized. One VCSEL has $x = 6\%$ and $N_p = 25$ top Bragg mirror pairs. It exhibits a minimum I_{th} at around 20°C. At this point, the cavity resonance and gain peak approximately match. By reducing x to 4.5% and 4%, the point of minimum I_{th} is shifted to around 55 and 75°C, respectively. The latter has a minimum I_{th} within the operating range of the atomic clocks (see Sect. 2.3.4). The magnitude of I_{th} depends on x, N_p, the active diameter D_a as well as the presence or absence of a surface grating. The latter will be explained in detail in Sect. 4.4. The main concern of the experiments summarized in Fig. 4.9 (left) was to optimize the point of minimum I_{th}. Therefore, QWs with 4% In content have been employed in the last generations of atomic clock VCSELs towards the end of the research performed for this dissertation. The three VCSELs are selected carefully to emit at wavelengths close to the target wavelength of 894.6 nm at 20°C. However, the devices have different internal heating due to different D_a and different bias currents. Therefore, to compare the emission wavelength of the devices, several spectra for different bias currents have been measured for each of the three VCSELs. Then, the wavelength of each device has been extrapolated to the hypothetical value at 0 mW dissipated power, as shown in Fig. 4.9 (right). The three VCSELs have approximately the same hypothetical emission wavelength of about 893.4 nm at 0 mW dissipated power and $T = 20°C$.

4.3 Single-Mode Emission

Due to their short cavity combined with the finite width of the stop-band of their DBRs, VCSELs are inherently longitudinal single-mode. Therefore, the term "mode" denotes a transverse mode in the context of VCSELs. Single-mode VCSELs are thus lasers which are emitting in one transverse mode. For atomic clock VCSELs, the single mode is the fundamental LP_{01} transverse mode[2]. There are several definitions of single-mode

---
[2]The various transverse modes in VCSELs are denoted by LP_{lp}, where LP stands for linearly polarized, the integer index $l \geq 0$ describes the azimuthal mode dependence with $2l$ denoting the number of intensity maxima on the circumference, and the integer index $p \geq 1$ characterizes the radial mode dependence and denotes the number of intensity maxima of the mode on the radius. LP_{01} is the lowest or fundamental transverse mode and has an approximately Gaussian-shaped beam profile.

4.3 Single-Mode Emission

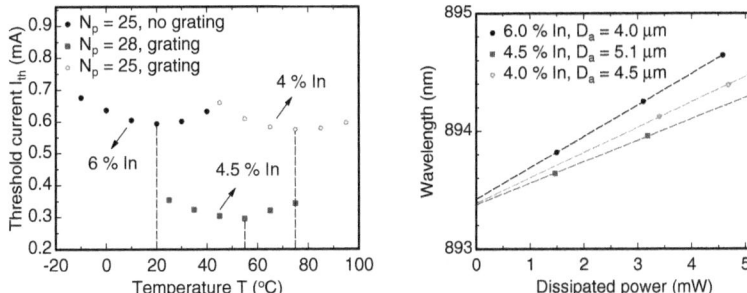

Figure 4.9: Experimental threshold current dependence on temperature for three atomic clock VCSELs with QWs having 6%, 4.5% and 4% indium content (left). Emission wavelength of the same VCSELs against dissipated power at $T = 20°C$, where the lines are linear fits (right). The VCSELs differ also in other design parameters such as the number of top mirror pairs N_p, the active diameter D_a, or the presence of surface grating.

VCSELs. The definition required by the MAC-TFC specifications and employed in this dissertation requests an SMSR of 20 dB, defined as the peak-to-peak difference between the fundamental mode (i.e., LP_{01}) and the first higher-order (i.e., LP_{11}) mode in the emission spectrum. For oxide-confined VCSELs, single-mode emission can be achieved by small oxide apertures, in analogy to single-mode fibers with small core diameters. VCSELs with small oxide diameters below 4 µm commonly show single-mode oscillation. However, such devices have the drawbacks of increased ohmic resistances and reduced lifetimes owing to higher current densities and possibly increased internal temperatures resulting from higher thermal and electrical resistances. Oxide-confined VCSELs with larger active diameters inherently show multi-mode oscillation. However, they can be forced to oscillate only at the fundamental mode by, e.g., etching a shallow surface relief in the upper DBR of a regular VCSEL structure [145], as depicted in Fig. 4.10 (left). An annular etch of the laser outcoupling facet lowers the effective mirror reflectivity particularly for higher-order modes, which show higher optical intensities outside the device center. The resulting differences in threshold gains strongly favor the fundamental mode.

In a regular VCSEL structure, a topmost GaAs cap layer is designed to provide a reflection at the semiconductor–air interface that is in-phase with the reflections at the interfaces in the DBR below. This means that the cap layer thickness is optimized for the highest reflectivity of the upper DBR and hence the lowest threshold gain. For a better understanding of the principle of the surface relief, the impact of the thickness of the cap layer on its reflectivity and consequently on the cavity losses of a VCSEL is

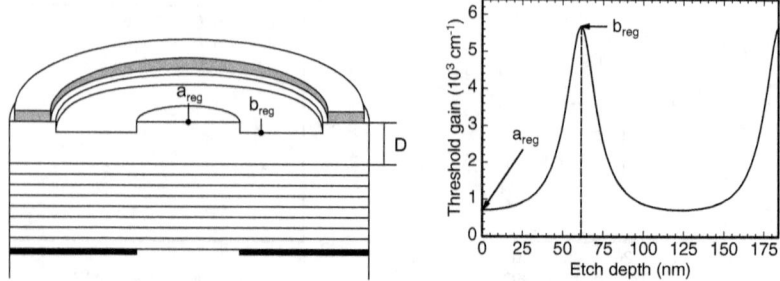

Figure 4.10: Schematic drawing of the upper DBR of a VCSEL structure with a regular surface relief (left), and calculated material threshold gain versus etch depth of the cap layer starting from a regular VCSEL structure (right). Points a_{reg} and b_{reg} indicate the unetched circular and the etched ring-shaped regions on the laser outcoupling facet, respectively.

depicted in Fig. 4.10 (right) in terms of the threshold gain, which is calculated using the transfer-matrix method [66]. The VCSEL layer structure[3] is the one shown in App. E, but with a cap layer thickness $t_{cap} = D = 246$ nm which corresponds to an optical path length of one emission wavelength in the material, $\lambda_{mat} = \lambda/\bar{n}$, with an emission wavelength $\lambda = 894.6$ nm and a refractive index of the cap layer $\bar{n} \approx 3.6$. For an etch depth of 0 nm (i.e., region a_{reg} in Fig. 4.10), a minimum threshold gain is achieved. For an etch depth of 61.5 nm (i.e., region b_{reg} in Fig. 4.10) which corresponds to $\lambda_{mat}/4$, a narrow local maximum of the threshold gain is observed. Therefore, applying an annular etch with a quarter-wave depth in the laser output facet lowers the mirror reflectivity and correspondingly increases the threshold gain outside the device center to be approximately a factor of seven larger than the threshold gain at the center. The resulting difference in threshold gains strongly enhances the fundamental-mode emission. The disadvantage of this technique is the very precise depth of a quarter-wave to which the area outside the relief (i.e., region b_{reg}) has to be etched.

To overcome this problem, a more advanced approach was developed in which the thickness of the topmost GaAs cap layer of the upper DBR in a regular structure is decreased by $\lambda_{mat}/4$, as depicted in Fig. 4.11 (left). The reflection at the semiconductor–air interface is anti-phase to the reflections at the other interfaces of the DBR. In other words, the cap layer thickness is optimized for the lowest reflectivity of the upper DBR. This structure is called an inverse structure and the resulting device is called *inverted relief VCSEL*. Here, the anti-phase layer is removed only in the center of the outcoupling facet, as depicted

---
[3]The VCSEL layer structure illustrated in App. E will be explained in detail in Sect. 4.5.

in Fig. 4.11 (left), and consequently the threshold gain for the fundamental mode is most strongly decreased. The dependence of the threshold gain of a VCSEL on the etch depth of the cap layer is shown in Fig. 4.11 (right). For an etch depth of 0 nm (i.e., region a_{inv} in Fig. 4.11), the threshold gain of the inverse structure is maximum. For an etch depth of 61.5 nm (i.e., region b_{inv} in Fig. 4.11), a broad minimum of the threshold gain can be seen. Hence, the inverted surface relief shows the advantage of relaxed dependence of the threshold gain on the etch depth compared to the regular surface relief. Technology-wise, this approach requires a less precise etch depth control and has been demonstrated with active diameters of 6 μm [146] and 5 μm [147] to provide maximum single-mode output powers of up to 6.3 mW and 6 mW, respectively, with SMSR > 30 dB. The relief diameters are 3.3 and 2.5 μm, respectively. It turns out that the optimum ratio of the diameters of relief and oxide aperture is equal to or slightly larger than 0.5 [146, 147]. Compared to a regular relief, it is clearly much easier to start from an inverse structure and etch the cap layer just in the area of the relief by about 61.5 nm to reach the broad valley between the peaks of the threshold gain (see Fig. 4.11 (right)) than to start etching from a regular structure and try to hit the narrow peak of the threshold gain outside of the relief (see Fig. 4.10 (right)).

Alternative approaches to single-mode oxide-confined VCSELs with large active diameters are summarized in Ref. [148]. These include i) using a combination of selective oxidation and ion implementation [149], ii) using an extended inner cavity [150], iii) using a small surface metal aperture [122], or iv) using a disordered region in the upper DBR by, e.g., Zn-diffusion [151]. Among all the approaches to single-mode VCSELs, the surface relief is the most advantageous technique, since it does not impair the electrical or thermal characteristics of regular VCSELs and yields high single-mode output powers with low threshold currents, high efficiencies as well as low series resistances. Moreover, it does not require a major modification of the regular VCSEL structure, facilitating manufacturability with low additional cost especially for large production volumes.

4.4 Polarization Control

As discussed in Sect. 3.6, there is no general polarization selection rule that could be identified for standard oxide-confined VCSELs. Switches between the main polarization directions can occur with changes in the operating temperature or current [11, 105–107]. Polarization stability of the VCSEL emission is of great interest for many applications including atomic clocks. Due to the birefringence introduced by the electro-optic and elasto-optic effects in VCSELs, the emission frequencies of the two polarization modes differ by up to 80 GHz [45], exceeding the spectral linewidths of Cs D_1 absorption lines

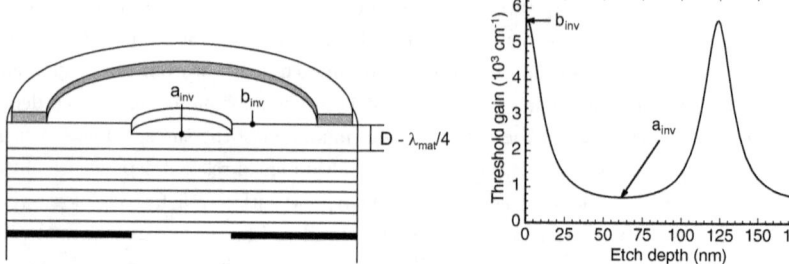

Figure 4.11: Schematic drawing of the top DBR of a VCSEL structure with an inverted surface relief (left), and calculated material threshold gain versus etch depth of the cap layer starting from an inverse VCSEL structure (right). Points a_{inv} and b_{inv} indicate the etched circular and the unetched ring-shaped regions on the laser outcoupling facet, respectively.

with a FWHM of approximately 390 MHz at 80°C [43]. In the case of a polarization switch, the spectral overlap of the atomic absorption lines and the laser modulation sidebands would be lost and the stable operation of the atomic clock could not be ensured anymore.

Over the last two decades, several research attempts were made to find reliable techniques and solutions to stabilize the polarization of VCSELs. The major approaches to polarization control of VCSELs can be roughly classified in four categories, namely use of i) asymmetric resonators, ii) polarization dependent gain, iii) external optical feedback or iv) polarization dependent mirrors [12].

Introducing a transverse asymmetry in the VCSEL cavity was the most popular approach. First attempts of this technique were realized by dumbbell-shaped [152], elliptical [72] or rectangular [153] mesas. Moreover, a tilted mesa structure [154] and a zigzag pattern at two opposing sidewalls of a rectangular mesa [155] were tried. However, in all cases the polarization control created by such transverse anisotropies remained rather weak.

The gain of QWs grown on (100)-oriented substrates is equal for all polarization directions in the plane of QWs if the direction of current flow is orthogonal to the QWs. However, the gain isotropy is broken as soon as the direction of the current flow is changed or when the QWs are grown on substrates with higher indices [12]. Anisotropic gain has been realized by VCSELs grown on GaAs (311)A [156, 157] and later on (311)B [158, 159] substrates[4], where the former shows problems in the n-doping of layers with high aluminium content [160]. With this technique for polarization control, VCSELs are polarized along the

---

[4]The symbols 'A' and 'B' indicate whether the surface layer contains gallium or arsenic atoms, respectively.

4.4 Polarization Control

[$\bar{2}$33] crystal axis with a maximum orthogonal polarization suppression ratio (OPSR) of 30 dB[5] [158]. Apart from the higher-index substrates being more expensive and having a lower quality than standard material, the epitaxial growth and also the processing is more difficult. For instance, the temperature for wet-thermal oxidation must be raised to 480°C, while usually temperatures around 400°C are chosen for a better control of the oxidation rate [161]. Nevertheless, single-mode VCSELs have been fabricated and show a stable polarization under CW operation with a peak-to-peak difference between the dominant and the suppressed polarization modes reaching 30 dB. However, this value degraded down to 11 dB under 5 GHz sinusoidal modulation [161].

External optical feedback has been verified to be a powerful approach for polarization control of VCSELs [162-164]. However, this technique requires relatively bulky and expensive optical components as well as careful assembly and mounting which abrogate the major favorable properties of VCSELs like compactness and very low price of much less than one euro for large production volumes. Besides anisotropic cavities, anisotropic gain and optical feedback, mirrors with a polarization-dependent reflectivity are another possibility for polarization control of VCSELs. In early attempts, several combinations of a DBR with a metallic or semiconductor grating have been tried to realize polarization-dependent mirrors [165–168]. However, none of these attempts was satisfactory due to technological restrictions [165], significant manufacturing complexity [166], incorporation difficulty with a VCSEL structure [167], or inferior device performance [168]. Instead of a grating, a polarization-dependent reflectivity has been accomplished by etching a shallow elliptic surface relief in the upper DBR [169, 170], but the polarization control created by this method is rather insufficient. Photonic crystals have been also employed in the upper DBR to realize polarization-stable VCSELs [171]. However, such VCSELs show moderate output powers and weak polarization control. Moreover, this approach is not straightforward. In 2002 the concept of a grating was proposed again to make a polarization-dependent mirror, but this time it was realized monolithically by etching a shallow surface grating in the upper DBR of the VCSEL structure. This technique was proposed by Pierluigi Debernardi [172] and it differs from the above-stated attempts in several respects. First, it is a pure semiconductor–air grating which has small absorption and a high refractive index contrast. It can produce high polarization-dependent reflectivity, but does not require a major modification of the VCSEL structure. Owing to the shallow semiconductor ridges, the diffraction and absorption losses can be kept acceptable. The realization of such monolithically integrated surface gratings is demonstrated in Fig. 4.12 by a scanning electron micrograph of a VCSEL with a surface grating etched in the topmost layer of the upper DBR.

---

[5]The suppressed orthogonally polarized mode is along the [01$\bar{1}$] crystal axis.

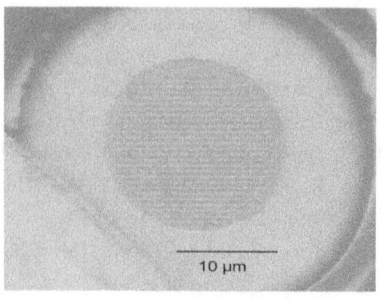

Figure 4.12: Scanning electron micrograph of a fully processed VCSEL incorporating a surface grating.

4.4.1 Concept of Surface Gratings for Polarization Control

Among all of the above-stated polarization control techniques, the monolithic incorporation of a linear semiconductor surface grating at the outcoupling facet was found to be most advantageous. The grating can be integrated in a standard VCSEL structure with very little additional processing effort and thus low cost. Namely, only the definition of the grating and its subsequent etching at the beginning of the fabrication sequence are needed, as will be illustrated and discussed in Sect. 4.6. For research purposes, the most suitable and flexible lithography technique to define the grating is electron-beam lithography. However, this technique might be too expensive for mass production of very-low-cost VCSELs. In these cases, nanoimprint lithography can be used [173].

The surface grating can be designed such that there is no penalty for laser characteristics like threshold current, differential quantum efficiency, circularly symmetric far-field emission profile, or high modulation bandwidth [12]. Insensitivity to optical feedback and external stress has also been proven. The results are documented in a series of publications which are summarized in [12, 174]. It is worth to note that the grating-based polarization control technique was commercialized very soon after its invention and is applied today in a large fraction of 850 nm wavelength VCSEL sensors in optical navigation devices like computer mice [113, 114]. In more detail, the following achievements have proven the success of the grating concept [174]:

- Demonstration of the first monolithic, truly polarization-stable VCSELs on regular GaAs substrates [175, 176].

- Excellent agreement between theoretical simulations and experimental results [177, 178].

- Highest reported quantity and yield of polarization-stable VCSELs on a given sample: 1374 VCSELs without any exception [174].

- Highest reported OPSRs: 26 dB [120] and 40 dB [179] when referring to the optical output powers and the spectral intensities, respectively.
- Highest reported linear dichroism (i.e., difference of material threshold gains of both polarization modes) in a VCSEL [180], specifically more than 40 cm^{-1}.
- First-ever demonstration of polarization control under high-frequency modulation [181], externally induced anisotropic strain [182], and optical feedback [183].
- Easy to adapt to different wavelength regimes, since the major grating parameters scale with the emission wavelength [174].

Therefore, the present work on VCSELs for atomic clocks also relies on such pure semiconductor–air surface gratings [80, 121].

4.4.2 Design of Surface Gratings

The design parameters of grating VCSELs in this work have been obtained by a vectorial, three-dimensional model based on the coupled-mode theory [13, 172]. The simulation results will be shown in Sect. 4.4.3. The polarizing effect of the surface grating originates from the difference in optical losses (or top mirror reflectivity) and thus threshold gains of modes polarized parallel or orthogonal to the grating lines. Hence the polarization mode with lowest threshold gain usually contributes most strongly to laser emission. With fundamental mode emission, there are only two polarization modes which are polarized parallel and orthogonal to the grating lines. The difference in the material threshold gains of the two polarization modes is commonly called *dichroism*. The larger is the dichroism, the higher is the stability of the laser polarization. It is convenient to characterize the strength of the polarization control by the *relative dichroism*

$$\mathrm{RD} = \frac{g_{\mathrm{orth}} - g_{\mathrm{para}}}{0.5(g_{\mathrm{orth}} + g_{\mathrm{para}})}, \qquad (4.28)$$

where g_{orth} and g_{para} are the material threshold gains of the two fundamental modes polarized orthogonal and parallel to the grating lines, respectively. The sign of RD indicates whether the dominant polarization of the emitted light of the VCSEL is parallel or orthogonal to the grating lines. If the threshold gain of the orthogonal mode is larger than that of the parallel mode, RD > 0 and the mode polarized parallel to the grating lines lases. Likewise, for RD < 0 one expects orthogonally polarized emission. As stated in Sect. 4.4.1, surface gratings are easy to adapt to different devices, since the major grating parameters scale with the emission wavelength [174].

The important grating parameters are indicated in the schematic drawing of Fig. 4.13, which depicts a surface grating etched in the topmost cap layer of the upper DBR of a

4 Design and Fabrication of VCSELs for Miniaturized Atomic Clocks

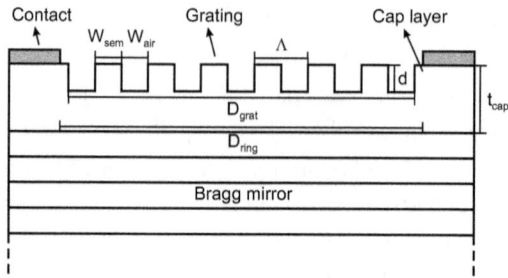

Figure 4.13: Schematic cross-section of the top part of a surface grating VCSEL indicating the important grating parameters.

VCSEL. Here d is the grating etch depth, Λ is the grating period, W_air and W_sem are the widths of grating groove and ridge, respectively, and t_cap is the thickness of the cap layer. D_grat and D_ring are the diameter of the circular area over which the grating pattern extends and that of the contact ring opening, respectively. Throughout this dissertation, the grating duty cycle, which is the ratio of the groove width and the grating period, is selected to be $W_\text{air}/\Lambda = 50\%$. Moreover, the grating lines are etched along the [011] crystal axis, which, resulting from the electro-optic effect, is one of the two preferred orthogonal polarization directions of standard GaAs VCSELs, as explained in Sect. 3.6. The grating is etched in a topmost GaAs cap layer with sufficient thickness $t_\text{cap} > d$, since exposing one of the AlGaAs layers to the ambient air strongly enhances the oxidation rate of the surface and reduces the reliability of the laser [184]. The thickness of the cap layer defines the relative longitudinal position of the surface grating with respect to the optical standing-wave field inside the resonator. This grating design parameter is found to be highly important and influences the polarization control and the overall performance of a grating VCSEL. Based on the longitudinal position of the surface grating, there are particularly two types of grating VCSELs. The first type is the *regular grating VCSEL* [175–177, 185], in which the grating is etched in a GaAs cap layer of a regular VCSEL structure described in Sect. 4.3. Here the thickness of the cap layer is optimized for the highest reflectivity of the upper DBR. For the second type of grating VCSEL, the thickness of the cap layer of a regular VCSEL structure is increased or decreased by $\lambda_\text{mat}/4$, leading to a minimum reflectivity of the upper DBR. One speaks of an *inverted grating VCSEL* [186]. If the thickness of the cap layer of a regular grating VCSEL is $D \geq \lambda_\text{mat}$, it can, for instance, be modified to be $D - \lambda_\text{mat}/4$ for an inverted grating VCSEL, as depicted in Fig. 4.14. As can be noticed, the grating pattern extends over the entire area of the outcoupling facet in order to provide the maximum transverse overlap between the grating and the laser modes.

4.4 Polarization Control

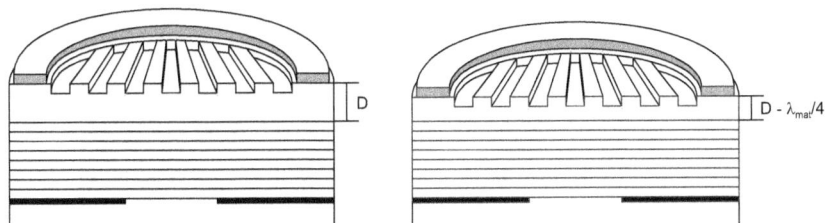

Figure 4.14: Schematic drawing of the upper DBR of a VCSEL structure into which a regular grating (left) and an inverted grating (right) are etched.

In order to avoid diffraction losses in air [121, 186], gratings with sub-emission-wavelength periods of 0.6 or 0.7 μm are employed in all grating VCSELs fabricated during the research done for this dissertation. Reducing the area of the surface grating to a diameter which is small enough to provide a transverse overlap between the surface grating and mainly the fundamental laser mode results in a favorable enhanced fundamental-mode emission as well as polarization-stable laser oscillation. This special type of grating is called *surface grating relief*. Similar to the full-aperture surface gratings, grating reliefs could be etched in a topmost GaAs cap layer of a regular VCSEL structure resulting in a regular grating relief VCSEL [175, 176], as depicted in Fig. 4.15 (left), or could be etched in a cap layer of an inverse VCSEL design resulting in an inverted grating relief VCSEL [120, 187, 188], as depicted in Fig. 4.15 (right). The latter technique is more favorable because of its much easier fabrication. By comparing the two structures in Fig. 4.15, for an inverted grating relief only some grooves have to be etched into the flat outcoupling facet instead of etching almost the complete facet except some grating ridges in the case of a regular grating relief. The second advantage of an inverted grating relief will be explained when the simulation results of surface grating VCSELs are introduced in the next sub-section. Different from the surface reliefs, $D_{\text{grat}}/D_{\text{a}}$ of the inverted grating relief devices should be in the range of 0.6 to 0.75 [120, 187, 188].

4.4.3 Simulations of Surface Grating VCSELs

Simulations of the layer structure of 894.6 nm atomic clock VCSELs shown in this section have been done by Pierluigi Debernardi using a fully vectorial, three-dimensional model based on coupled-mode theory [13, 172]. Despite of the multitude of non-idealities in real devices like non-circular geometry, the model has proven to very well predict the cold-cavity properties of VCSELs [177, 178]. Simulation results are the modal emission wavelengths, the cold-cavity optical losses of the individual polarization modes of all

4 Design and Fabrication of VCSELs for Miniaturized Atomic Clocks

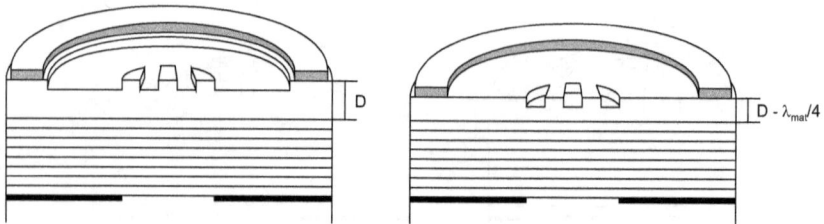

Figure 4.15: Schematic drawing of the upper DBR of a VCSEL structure with a regular grating relief (left) and an inverted grating relief (right).

transverse modes, as well as their field distributions. However, in the present work we are concerned only with the fundamental transverse mode and its two orthogonal polarization modes. With the model, one can simulate the material threshold gains of the two polarization modes needed to overcome the cold-cavity losses introduced partially by the surface grating.

The VCSEL layer structure for which simulations have been done is identical to the one stated in App. E, but with a cap layer thickness $t_{cap} = 3\lambda_{mat}/4$ and $\lambda_{mat}/2$ for inverse and regular VCSEL design, respectively[6]. Moreover, for the simulations of both designs, an active diameter of 4 µm, a grating period of 0.7 µm, and a duty cycle of 50% are assumed. The grating diameter is assumed to be equal to the inner diameter of the p-ring contact, i.e., $D_{grat} = D_{ring}$. The dependence of the material threshold gains of the two fundamental transverse modes polarized parallel and orthogonal to the grating lines of an inverted grating VCSEL are depicted in Fig. 4.16 (left) with grating depths d varied between 30 and 100 nm. The relative dichroism calculated from the threshold gains is also plotted in the figure. A negative relative dichroism is observed over the whole range of grating depths, which indicates that the VCSEL emission is polarized orthogonal to the grating lines. In order to have the lowest threshold gain and hence the minimum threshold current, d should be around 65 nm, which corresponds to about $\lambda_{mat}/4$.

The simulation results for a regular grating VCSEL with the same active diameter of 4 µm are depicted in Fig. 4.16 (right). For grating depths below 42 nm, the threshold gain

---
[6]The cap layer thickness t_{cap} for the inverse design is identical to what has been used for the calculation of threshold gains using the transfer-matrix method depicted in Fig. 4.11 (right). However, t_{cap} for the regular design is shorter by $\lambda_{mat}/2$ compared to Fig. 4.10 (right). Due to the periodicity of the phase of the reflection coefficient of the upper DBR when t_{cap} is increased by multiples of $\lambda_{mat}/2$, the associated simulated material threshold gains show only negligible increase caused by free-carrier absorption in such additional thickness of the cap layer. As well, relative dichroisms calculated from the material threshold gains stay almost unchanged due to almost equal increase of g_{orth} and g_{para} in (4.28).

4.4 Polarization Control

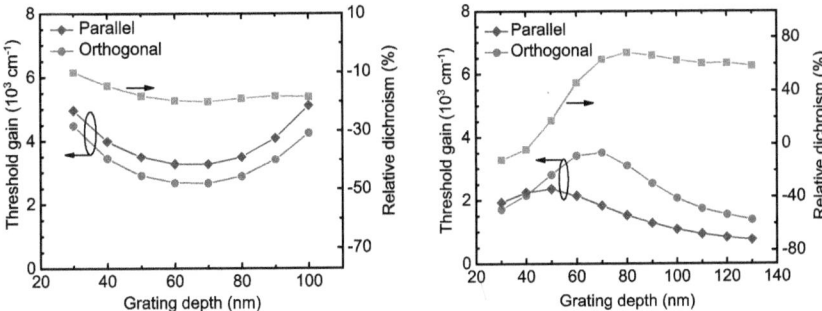

Figure 4.16: Simulated material threshold gains g of the two fundamental polarization modes (left axis) and the corresponding relative dichroism (right axis) calculated from the threshold gains (according to (4.28)) as a function of the grating depth for an inverted (left graph) and a regular (right graph) grating VCSEL. Both lasers have 4 µm active diameter and $N_p = 25$. The surface grating has 0.7 µm period and 50% duty cycle.

of the parallel polarized mode is larger than that of the orthogonal polarization. The situation is reverse for larger grating depths. Thus, the relative dichroism changes its sign from negative to positive at $d_c = 42$ nm, favoring the parallel polarization for $d > d_c$. Figure 4.16 (right) shows that the material threshold gain of the selected polarization mode is minimum at 120 to 130 nm grating depth, which corresponds to about $\lambda_{mat}/2$. The minimum threshold gain is approximately 30% of the corresponding minimum of the inverted grating VCSEL at $d = 60$ to 70 nm in Fig. 4.16 (left). Reduced threshold currents of regular grating VCSELs are thus expected and will be shown in the experimental part of this dissertation in Chap. 6.

Besides its much easier fabrication as explained in Sect. 4.4.2, an inverted grating relief is favored over a regular grating relief because of a second important advantage which was not explained in that sub-section. Like a regular surface relief, the regular grating relief requires a very exact etch depth of $\lambda_{mat}/4$ outside the relief to obtain the highest difference between threshold gains of the fundamental and the higher-order modes and therefore a strong mode control is achieved, as illustrated by Fig. 4.10. However, the grating depth is also fixed to $d = \lambda_{mat}/4 = 61.5$ nm and cannot be optimized independently to achieve the best polarization control by obtaining the highest relative dichroism (e.g., at $d = 80$ nm) or cannot be optimized to achieve the lowest material threshold gain and consequently the minimum threshold current (e.g., at $d = 130$ nm), as can be seen from Fig. 4.16 (right). These problems can be avoided with an inverted grating relief, by which the threshold

69

4 Design and Fabrication of VCSELs for Miniaturized Atomic Clocks

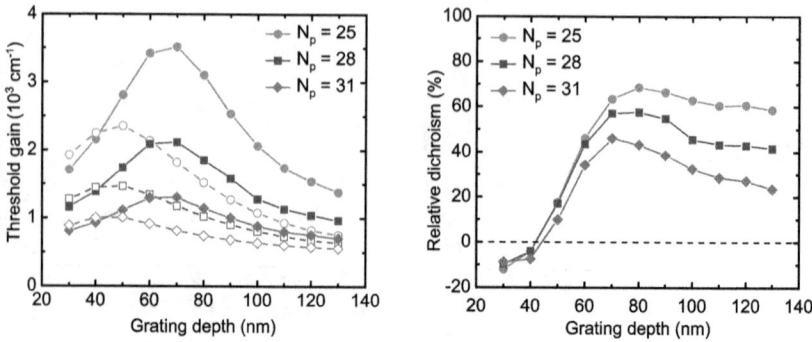

Figure 4.17: Simulated material threshold gains of the two fundamental polarization modes as a function of grating depth of the regular grating VCSEL structure from Fig. 4.16 (right) for different numbers of upper DBR pairs N_p (left) and corresponding relative dichroisms (right). The full and open symbols in the left graph represent the material threshold gains of the two fundamental modes polarized orthogonal and parallel to the grating lines, respectively.

gain of the higher-order modes is defined and fixed by the epitaxial surface. Therefore, the grating depth can be independently optimized to achieve a good polarization control and the lowest material threshold gain (e.g., at $d = 60$ to $70\,\text{nm}$), as can be seen from Fig. 4.16 (left). A major tradeoff in designing grating VCSELs are the material threshold gain and the relative dichroism achieved with different numbers of layer pairs of the outcoupling mirror. Figure 4.17 (left) illustrates the simulated material threshold gains of the two fundamental modes polarized parallel and orthogonal to the grating lines in dependence of the grating depth for different N_p of the regular grating VCSEL from Fig. 4.16 (right). With increasing N_p, the threshold gains decrease due to the increased reflectivity of the top DBR. On the other hand, there is a lower field intensity at the surface grating (i.e., the surface grating is more decoupled from the laser cavity) and its polarization control effect becomes thus weaker. This trend is reflected in the relative dichroism, which decreases with increasing N_p, as depicted in Fig. 4.17 (right). Polarization stability is a high priority for atomic clock VCSELs, to be ensured under all adverse operating conditions like high-frequency modulation or possible optical feedback in the clock microsystem. To be on the very safe side, N_p in this dissertation is limited to 28. A higher N_p might still provide sufficient polarization stability in practice, with the benefit of further reduced I_{th} and higher insensitivity to optical feedback. The available output power which decreases with higher N_p is not a major concern since the MAC does not require more than $100\,\mu\text{W}$ (see Sect. 2.3.4). At this point, it is not possible to identify a strict upper limit for N_p.

4.5 Layer Structure

The VCSEL layers are grown on n-doped (100)-oriented GaAs 2-inch wafers using solid-source molecular beam epitaxy (MBE). There is a highly n-doped GaAs contact layer above the GaAs substrate to allow n-contacting. The active region consists of three compressively strained 8 nm thick $In_xGa_{1-x}As$ QWs separated by 10 nm thick $Al_{0.27}Ga_{0.73}As$ QBs with $x = 6\%$, 4.5% or 4% indium content[7]. The QWs are placed in an antinode of the standing-wave pattern to achieve a good coupling between electrons and photons. The active region is sandwiched between two larger bandgap $Al_{0.47}Ga_{0.53}As$ cladding layers to form a one-wavelength-thick inner cavity. A highly p-doped 30 nm thick AlAs layer is grown directly above the top cladding layer at a node of the standing-wave pattern. Current confinement and weak optical index guiding are achieved by wet-thermal oxidation of the AlAs layer. The n-type bottom and p-type top DBRs consist of 38.5 Si-doped and 25 or 28 C-doped $Al_{0.90}Ga_{0.10}As/Al_{0.20}Ga_{0.80}As$ layer pairs. The DBRs are graded in doping concentration [189] and composition [190] to minimize the free-carrier absorption and decrease the electrical resistance. The structure has an extra topmost GaAs cap layer in which the surface grating for polarization control is etched. The thickness of the cap layer is $\lambda_{mat}/4$ or $\lambda_{mat}/2$ to construct either inverse or regular VCSEL structures, respectively. The detailed layer structures are shown in App. E.

An inherent problem of the available epitaxy machine is the thickness gradient of the VCSEL layers originating from spatially inhomogeneous growth. This causes variations by several tens of nanometers of the emission wavelength λ of the final fully-processed devices. VCSELs located at the center of the wafer show the maximum λ, while λ decreases for devices away from the center. Therefore, the yield of VCSELs with the target emission wavelength of $\lambda = 894.6$ nm is very small. Another yield issue are variations during the wet-thermal oxidation step by which the active aperture is produced. The oxidation rate is highly dependent on the Al content and the thickness of the AlAs layer. The resultant aperture size mainly determines if the laser is multi-mode or single-mode and even influences the SMSR. Only a fraction of the VCSELs with the target wavelength could satisfy all the requirements on the laser parameters stated in Sect. 2.3.4. This imposes an additional reduction of the yield of the MAC-suitable VCSELs. Nevertheless, commercial VCSELs grown by industrial-level epitaxy machines typically feature a one to two orders of magnitude smaller thickness gradient over a 3-inch wafer [191]. Hence, the yield from such epitaxy material is expected to be much larger.

---

[7]Based on x, the temperature dependence of I_{th} is changed. As illustrated in Sect. 4.2.6, it is found that by reducing x, the point of minimum I_{th} is shifted to higher temperatures. In particular, for $x = 4\%$, a minimum I_{th} within the operating range (i.e., $T = 65$ to $80°C$) of the atomic clocks is achieved. However, in this dissertation, experimental results of VCSELs with different values of x are presented.

4 Design and Fabrication of VCSELs for Miniaturized Atomic Clocks

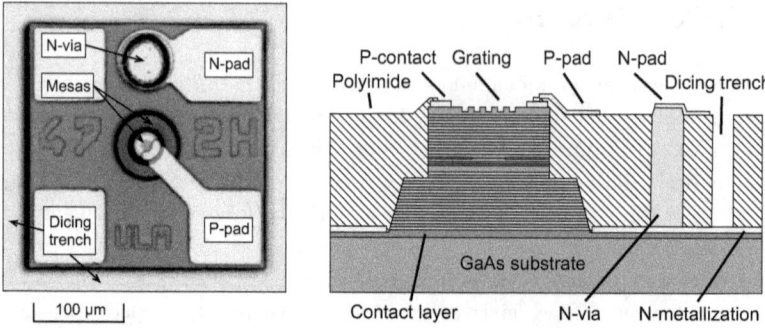

Figure 4.18: Optical micrograph of a fully processed flip-chip-bondable VCSEL chip to be incorporated into atomic clock microsystems (left) and a schematic cross-sectional drawing (right).

4.6 VCSEL Chip Design and Processing

In this section, the chip design of VCSEL devices used in the scope of this dissertation is illustrated and discussed. Technological fabrication processes applied to produce the VCSEL chips are presented as well.

4.6.1 Flip-Chip-Bondable Design

For the purpose of integration with the atomic clock microsystem, flip-chip-bondable VCSEL chips have been realized. As seen in Fig. 4.18 (left), the chip has a size of 300 × 300 μm^2 and can be tested on-wafer using coplanar microwave probes with a signal–ground (SG) configuration. Figure 4.18 (right) depicts a cross-sectional view of a VCSEL chip. The design involves increased processing complexity compared to regular VCSELs with substrate-side n-contacts (see Fig. 3.1). This originates from the necessity of etching mesas[8] down to the highly n-doped contact layer, applying several polyimide planarization layers containing dicing trenches as well as circular holes for n-vias, and electroplating of Au n-vias. As a last step, bondpads are evaporated.

---

[8] First, a mesa is etched in the p-type DBR to make the AlAs layer accessible for oxidation. This is called "p-mesa". Then a second mesa is etched in the n-type DBR to reach down to the highly n-doped contact layer. We speak of the "n-mesa".

72

4.6.2 VCSEL Processing

The fabrication of VCSELs is based on standard lithographic processes. The starting material is a GaAs wafer, on which the layer structure described in Sect. 4.5 is grown by solid-source MBE. Individual VCSEL devices are defined, where their order on the wafer is chosen by the design of the lithographic masks. Figure 4.19 schematically illustrates the processing steps essential to fabricate a flip-chip-bondable grating VCSEL. The framed numbers are used to refer to individual steps when discussing the processing procedure.

As a first step, electron-beam resist is deposited on the wafer surface and exposed using an electron-beam lithography machine to define the grating pattern and a circular edge whose purpose is to facilitate precise alignment of the subsequent mesa etch step (boxed 1, Fig. 4.19). The grating and the alignment edge are then etched using a solution of citric acid and hydrogen peroxide. To etch the p-mesa, the electron-beam resist is removed and a thicker photoresist is deposited and aligned to the predefined alignment edge (boxed 2, Fig. 4.19). The p-mesa is defined by a lithography step to be smaller than the alignment edge by 4 µm in diameter, which provides adequate alignment precision between the p-mesa (which predefines the oxide aperture) and the grating especially when using a grating relief, which usually has a diameter smaller than that of oxide aperture, as discussed in Sect. 4.4.2.

For etching the p-mesa (boxed 3, Fig. 4.19) either chemical wet-etching or dry-etching have been employed, leading to mesas with either inclined or steep side walls, respectively, as can be seen from Fig. 4.20. Chemical wet-etching is done using a sulfuric acid etch solution showing an etch rate of approximately 1.2 µm/min, where the etch depth can be controlled in situ by the color change between high and low aluminium containing layers. Dry-etching is done using the reactive-ion etching[9] (RIE) technique, where directed high-energy argon ions attack and etch the wafer surface. Material removal is mainly based on the mechanical impact at the semiconductor surface, leading to an anisotropic etching and steep side walls. Selectivity between the photoresist and the semiconductor material can be increased by employing $SiCl_4$, which can chemically etch the semiconductor material. The ratio between chemical and mechanical etch can be controlled by process parameters like pressure, gas composition and concentration, and RF power. A mesa etch rate of approximately 0.1 µm/min with a selectivity of up to 1:20 between photoresist and semiconductor is achieved. Next the AlAs layer is wet-thermally oxidized to form a current aperture (boxed 4, Fig. 4.19). This is done in a three-zone furnace in humid ambiance at a temperature range of 370 to 400°C. The chemical reaction

$$2AlAs + 3H_2O \rightarrow Al_2O_3 + 2AsH_3 \qquad (4.29)$$

---
[9]Oxford Instruments, model Plasmalab System100.

4 Design and Fabrication of VCSELs for Miniaturized Atomic Clocks

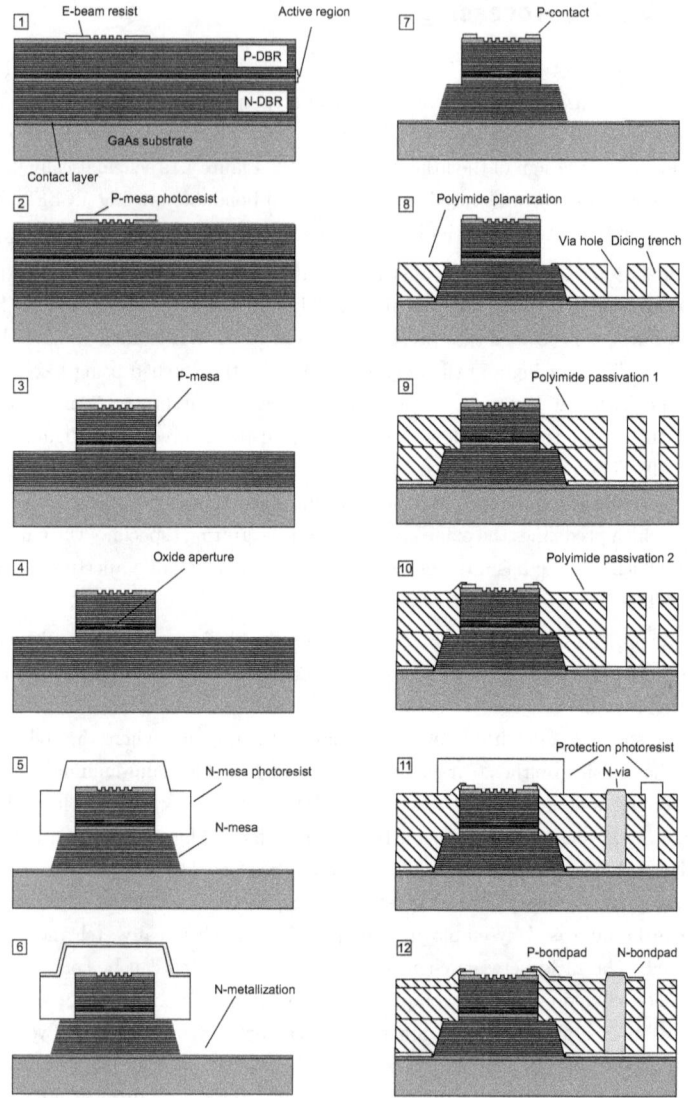

Figure 4.19: Illustration of the fabrication process for flip-chip-bondable VCSELs with surface grating.

4.6 VCSEL Chip Design and Processing

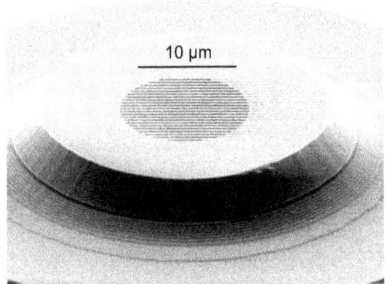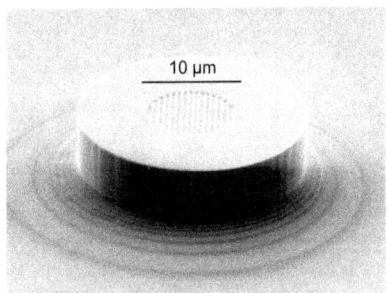

Figure 4.20: Scanning electron micrographs of a mesa after chemical wet-etching showing tilted side walls (left) and of a mesa after RIE dry-etching showing almost vertical side walls (right).

occurs and transfers the semiconductor to dielectric polycrystalline Al_2O_3 [192], where nitrogen is used as inert carrier gas for the water vapor. The resulting oxide aperture can be inspected by illuminating the structure using infrared light, which can penetrate the upper DBR. Owing to the difference in refractive index, the reflectivity from oxidized and non-oxidized areas differs. A charge-coupled device (CCD) camera is utilized to detect the reflected light which gives an image of the oxide aperture as shown in Fig. 4.21. The aperture has usually a non-circular or rhombic shape, since the oxidation rate is slightly faster along the $\langle 001 \rangle$ crystal axes compared to the $\langle 011 \rangle$ axes [193].

To make the n-contact accessible from the epitaxial side of the wafer, the p-mesa is protected by a thick photoresist and chemically wet-etched until reaching the highly n-doped contact layer ($\boxed{5}$, Fig. 4.19). Chemical wet-etching is favorable for this process, since the n-metallization can then be deposited on the same photoresist ($\boxed{6}$, Fig. 4.19). The etching undercut of the structured n-mesa ameliorates the lift-off process of the photoresist. For the n-metallization, an alloying contact consisting of Ge-Au-Ni-Au is used. On top of the p-type DBR a Ti-Pt-Au ring-shaped tunnel contact is evaporated ($\boxed{7}$, Fig. 4.19). The p-contact aperture is made larger (by $\approx 10\,\mu m$) than the active diameter to prevent clipping of the output beam at the metal. In order to overcome the large topography (7 to $8\,\mu m$), a planarization layer of thick polyimide is deposited and cured ($\boxed{8}$, Fig. 4.19). Next, two passivations are deposited using thick and thin polyimide layers ($\boxed{9}$ and $\boxed{10}$, Fig. 4.19). Only the second passivation layer overlaps with the p-mesa top, where thin polyimide is utilized to avoid an accumulated thick layer above the p-mesa top. Such a thick layer can cut the thin evaporated metal that will be applied in the following bondpad lithography step causing an electrical open circuit and preventing current flow to the laser active region. Both planarization as well as passivation layers

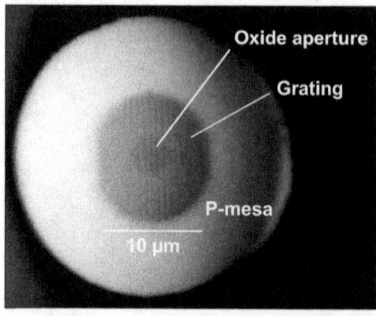

Figure 4.21: Infrared microscope image of an oxidized grating VCSEL.

contain cleaving trenches as well as circular holes for the n-vias, while device numbers are only imprinted into the thin passivation layer. The Au n-vias are then electroplated to reach the top of the polyimide stack of layers after protecting mesa and cleaving trenches with photoresist ([11], Fig. 4.19).

Finally, bondpads are structured on top of the polyimide and overlap with both p-ring contact and n-vias ([12], Fig. 4.19). The bondpads consist of Ni-Au and have $80 \times 80\ \mu m^2$ size, so electrical contacts can be easily made using coplanar microwave probes with a SG configuration. Since the wafer back side is unprocessed, substrate thinning down to approximately 200 μm is carried out to facilitate cleaving of solitary VCSEL chips with $300 \times 300\ \mu m^2$ size, as depicted in Fig. 4.18 (left). As will be explained in Chap. 7, some atomic clock VCSELs are packaged in TO-cans for convenient testing. Therefore, the bondpads have to be wire-bondable. This is achieved by applying a final annealing step at 330°C. To avoid cracks on the bondpads, it is necessary that the preceding polyimide passivation step is hard-baked at slightly higher temperature (e.g., 350°C). The lower-temperature bondpad annealing process would not thus cause further shrinkage of the subjacent polyimide layers, providing a crack-free transition of the bondpads on the mesa edges. Polyimide thicknesses of about 5.5 μm were measured (using a profilometer[10]) for such wire-bondable devices. Earlier fabricated VCSELs at the beginning of the research performed for this dissertation used to be not wire-bondable. The processing procedure for such devices does not include a final annealing process, moreover the final polyimide layer was hard-baked at slightly lower temperature (e.g., 300°C) resulting in quite higher polyimide thicknesses of about 7.5 μm. A detailed technological processing procedure can be found in App. D.1.

An alternative and less complicated processing procedure is described in detail in App. D.2, which applies one mesa etching step and a thin polyimide layer. For a thin polyimide

---

[10]Tencor, model Alpha Step 100.

4.6 VCSEL Chip Design and Processing

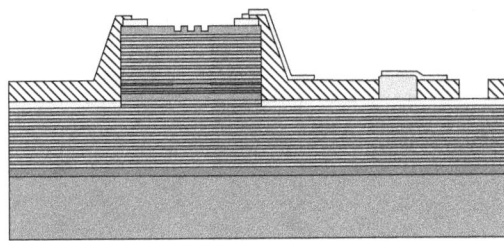

Figure 4.22: Schematic drawing of a flip-chip-bondable VCSEL chip like in Fig. 4.18 but with a thin polyimide layer and one mesa etching step down to the first layer pairs of the n-type DBR.

VCSEL, the highly n-doped GaAs contact layer indicated in Fig. 4.18 (right) is obsolete. The mesa etching step has to reach, at least, the first layer pair of the n-type DBR. However, the thickness gradient of the VCSEL layers at different wafer positions from its center (explained earlier in Sect. 4.5) causes variations of the etch depth. For instance, if a mesa at the wafer center is etched down to the first layer pair of the n-type DBR, the second layer pair is already reached at a position far away from the center. This can also cause variations of the n-contact resistance. Figure 4.22 schematically depicts a thin polyimide VCSEL. Polyimide thicknesses of about 1.5 μm were measured for such devices. The purpose and advantage of such devices will be discussed in Sect. 6.3.

Definition and Etching Process of the Grating

It was addressed in Sect. 4.4.2 that sub-emission-wavelength grating periods are more favorable than grating periods similar or even larger than the emission wavelength. Owing to its high resolution, electron-beam lithography is employed to define the gratings on all wafers processed during the research performed in this dissertation[11]. The resolution in electron-beam lithography is not limited by diffraction but by electron scattering. When the accelerated electrons enter the resist, they are scattered at the polymer molecules of the resist. This scattering generates slow secondary electrons with low energies between 2 and 50 eV [194]. The propagation of these secondary electrons inside the resist leads to a broadening of the electron beam by approximately 10 nm, which limits the maximum achievable resolution to about $\beta_0 = 25$ nm [194]. The primary electrons propagate also inside the resist under small scattering angles. Those are called forward-scattered electrons and they usually pass through the entire thickness of the resist causing much larger

---
[11]The electron-beam lithography system used throughout this dissertation was manufactured by Leica Microsystems GmbH, model EBPG 5 HR.

4 Design and Fabrication of VCSELs for Miniaturized Atomic Clocks

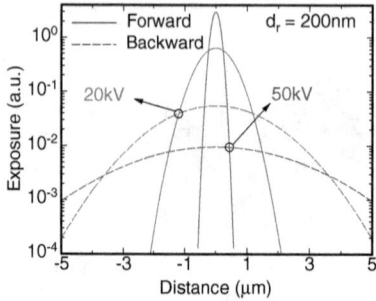

Figure 4.23: Exposure distributions by forward- (solid) and backward-scattered (dashed) electrons for a resist thicknesses $d_r = 200$ nm and acceleration voltages of 20 kV (green) and 50 kV (blue). β_f, β_b and η_E are set to 180 nm, 3.28 µm and 1.07, respectively.

broadening of the electron beam than the range which the secondary (low-energetic) electrons travel. The beam broadening due to the forward scattering decreases with increased acceleration voltage and decreased resist thickness and can be empirically given by [194]

$$\alpha_f = 0.9 \times 10^{18} \left(\frac{d_r}{V_{acc}}\right)^{1.5} = 0.9 \left(\frac{d'_r}{V'_{acc}}\right)^{1.5}, \quad (4.30)$$

with α_f as the electron-beam broadening factor, $d'_r = 10^9 d_r$ as the resist thickness in nanometers, and $V'_{acc} = 10^{-3} V_{acc}$ as the acceleration voltage in kilovolts. The thickness of the resist used for electron-beam lithography is 200 nm, which results in a broadening factor $\alpha_f = 7.2$ when 50 kV acceleration voltage is applied. Consequently, the electron-beam width is increased from $\beta_0 = 25$ nm (which is typical for the employed electron-beam lithography system) to $\beta_f = \alpha_f \cdot \beta_0 = 180$ nm. As the electrons propagate deeper through the resist and even through the sample, more and more of them are scattered under larger angles and eventually even backwards into the resist [194]. These backscattered electrons contribute to the exposure of the resist, too. This mechanism is called the *proximity effect*. The distance an electron can travel before losing all its energy primarily depends on the acceleration voltage and the semiconductor material of the wafer. In the case of GaAs and an acceleration voltage of 50 kV, this distance is more than 10 µm [194]. Besides the primary and slow secondary electrons, so-called fast secondary electrons are generated in the exposure process having a higher energy in the order of 1 keV. They contribute to the proximity effect in the order of some tenths of a micrometer [194].

The beam broadening caused by forward- and backward-scattered electrons depends on several parameters such as acceleration voltage V_{acc} and resist thickness d_r. The exposure distribution can be approximated by a double-Gaussian function describing the exposure at a radial position r from the (infinitely thin) primary beam as [195, 196]

$$f_e(r) = k_n \left[\exp\left(\frac{-r^2}{\beta_f^2}\right) + \frac{\eta_E \beta_f^2}{\beta_b^2} \exp\left(\frac{-r^2}{\beta_b^2}\right) \right], \quad (4.31)$$

4.6 VCSEL Chip Design and Processing

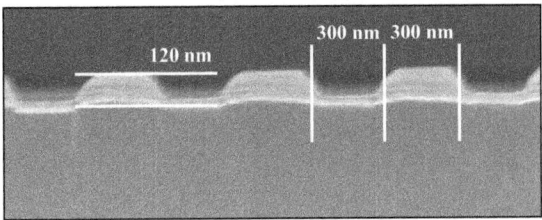

Figure 4.24: SEM micrograph of cross-sectional view of a surface grating which has 120 nm etch depth, 0.6 μm period, and 50% duty cycle. .

with k_n as a normalization factor, β_f and β_b are the beam widths[12] caused by the forward- and backward-scattered electrons, respectively, and the back-scattering coefficient η_E is the ratio of the resist exposure by backward-scattered electrons to that by forward-scattered electrons.

All parameters appearing in (4.31) are dependent on the depth $\bar{z}$ of the incident beam in the resist (where $\bar{z} \leq d_r$). β_f increases with increased $\bar{z}$ and decreased electron energy $E_e = qV_{\text{acc}}$. β_b increases with increased E_e, but its dependence on $\bar{z}$ is very weak. η_E is roughly independent of d_r and E_e [195]. The distributions of resist exposure by forward- and backward-scattered electrons are shown in Fig. 4.23. Distribution parameters β_f, β_b and η_E are taken from [196] and [197] for $V_{\text{acc}} = 20$ and $50\,\text{kV}$, respectively. The normalization factor k_n is selected to satisfy the condition

$$\int_{-\infty}^{\infty} f_e(r) \mathrm{d}r = 1. \tag{4.32}$$

For grating definition, when exposing one of the grating lines, the backward-scattered electrons transfer energy to all neighboring lines within the grating structure and thus partially contribute to their exposure. A high acceleration voltage of 50 kV is necessary to reduce β_f. This also increases β_b, but at the same time decreases the peak of the exposure distribution of backward-scattered electrons (quantified by $\eta_E \beta_f^2 / \beta_b^2$), as illustrated in Fig. 4.23. An efficient technique to eliminate the proximity effect is to apply equalizing reverse exposure to compensate the background dose caused by backscattered electrons [198]. However, in this work no correcting exposure was applied. Instead, the primary exposure dose (electrons per area) has been optimized so that the dose of the forward-scattered electrons is larger than the clearing threshold of the resist, while the exposure dose caused by the backward-scattered electrons is less than the resist clearing threshold.

---
[12]The electron-beam widths β_f and β_b are assumed to be the half width at $1/e$ of the exposure peak, where $e \approx 2.71828$ is the Euler constant.

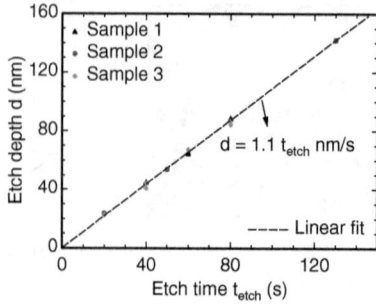

Figure 4.25: Measured grating depths of 18 VCSELs located on three nominally identical samples from different epitaxial runs. Each device is etched for a different time period depending on its position within the same sample. The linear fit (dashed line) determines the etch rate.

For optimizing the results of the electron-beam lithography, dose test wafers have been exposed, etched and analyzed. For the latter, an atomic force microscope[13] (AFM) and scanning electron microscope[14] (SEM) have been used during the research performed for this dissertation. The AFM is primarily used to measure the grating etch depth accurately. Due to the lateral extension of its tip and due to the limited number of data points which can be recorded along one line, AFM is not sufficiently accurate to analyze the transverse profile of the grating structure. Instead, an SEM is utilized. After defining the grating lithographically, it is transferred into the semiconductor by etching. The etch depths required for the grating fabrication are in the range of 60 to 125 nm. The SEM micrograph shown in Fig. 4.24 illustrates a cross-sectional view of a surface grating which is chemically wet-etched in the upper DBR. The tilted side walls of the grating ridges are due to the isotropic wet-etching behavior, where both vertical and lateral etch rates are comparable.

As mentioned earlier in this section, a citric acid solution is utilized for etching surface gratings. The solution is prepared by dissolving monohydrated citric acid ($C_6H_8O_7 \cdot H_2O$) in deionized water at a ratio of 1 g citric acid to 1 ml water. Afterwards, the citric acid solution is mixed with a solution of deionized water and hydrogen peroxide with a portion of 33% of the latter. The ratio of the citric acid solution to the hydrogen peroxide solution is 30:1 in volume. This mixing ratio results in a reaction-limited etching solution with which the etch rate is linearly proportional to the etch time, unaffected by stirring of the solution during etching [199]. Figure 4.25 depicts the grating etch rate of 18 VCSELs from three nominally identical samples[15] fabricated by different epitaxial growth runs. The different grating depths d at different sections within the same sample are achieved by applying different etching times. With a linear dependence on elapsed time and a

---
[13]Digital Instruments Corp., model D3100SPM.
[14]Carl Zeiss Microscopy GmbH, model LEO 982.
[15]Six VCSELs from each sample.

4.6 VCSEL Chip Design and Processing

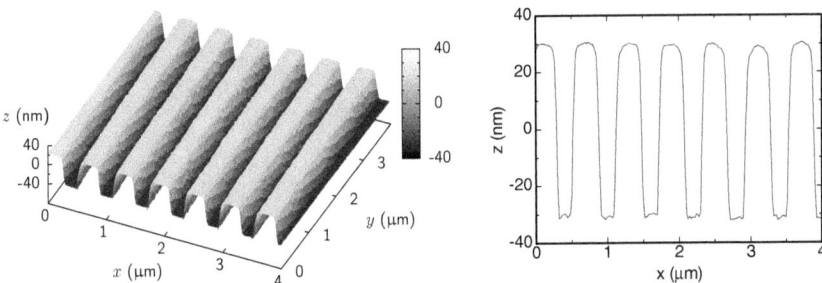

Figure 4.26: Surface profile (left) and cross-section (right) of a part of a chemically wet-etched grating measured by AFM. The grating has an etch depth of 60 nm and a grating period of 0.6 µm.

low etch rate of approximately 1.1 nm/s, the above citric acid solution is an excellent candidate for etching surface gratings with nanometer precision. The grating depth can be controlled quite precisely by the elapsed time. Moreover, the homogeneity of etch depths within different sections of the same sample and the reproducibility in different samples using identical etching times are obvious. The grating depths are measured using an AFM. A surface profile and a cross-section of a part of surface grating with 0.6 µm period and 60 nm depth are presented in Fig. 4.26.

Chapter 5

Experimental Characterization of Atomic Clock VCSELs

Flip-chip-bondable inverted grating VCSELs emitting at 894.6 nm for cesium-based atomic clocks have been successfully fabricated using the fabrication process explained in Sect. 4.6. In this chapter, static and dynamic characteristics of atomic clock VCSELs with InGaAs QWs having indium contents of $x = 4$, 4.5 or 6% are going to be presented. The p-type upper DBR has either $N_p = 25$ or 28 layer pairs. The inverted gratings have either 60 or 70 nm (i.e., approximately quarter-wave) depth and 0.6 or 0.7 µm period. The emission of these devices is always polarized orthogonal to the grating lines, which is consistent with the simulation results shown in Sect. 4.4.3. Most of the atomic clock VCSELs fabricated during the research performed for this dissertation employed inverted gratings for polarization control. However, atomic clock VCSELs employing inverted grating reliefs or regular gratings have been fabricated as well, and will be presented along with their advantages and drawbacks compared to inverted grating devices in the next chapter.

5.1 Static Characteristics

The target emission wavelength of 894.6 nm is the key selection parameter of VCSELs for cesium-based miniaturized atomic clocks. As stated in Sect. 2.3.4, the emission wavelength should be obtained at elevated ambient temperature (e.g., $T = 65$ to 80°C) and moderately high bias currents to guarantee a low power consumption and sufficient modulation bandwidth.

5 Experimental Characterization of Atomic Clock VCSELs

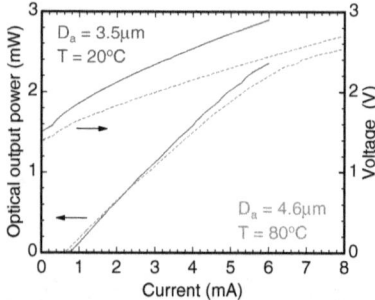

Figure 5.1: Operation characteristics of two 894.6 nm inverted grating VCSELs with active diameters of 3.5 µm (full lines) and 4.6 µm (dashed lines) at $T = 20$ and $80°C$ substrate temperatures, respectively. The surface grating has 70 nm depth and 0.6 µm period. Both lasers have $x = 4\%$ indium content in the QWs and $N_p = 25$ top mirror pairs.

5.1.1 Operation Characteristics and Emission Spectra

Figure 5.1 shows the LIV characteristics of two atomic clock VCSELs with $D_a = 3.5$ and 4.6 µm active diameter at $T = 20$ and $80°C$ substrate temperature, respectively. Their threshold currents are less than 1 mA. The device with $D_a = 4.6$ µm has a smaller threshold current of approximately 0.65 mA despite its larger oxide aperture and its higher substrate temperature. This is due to the optimized indium content in the InGaAs QWs which leads to minimum threshold currents within the operating temperature range of the atomic clock (e.g., $T = 65$ to $80°C$), as was explained in Sect. 4.2.6.

The CW spectrum of the VCSEL with $D_a = 3.5$ µm at $I = 1.6$ mA bias current and $T = 20°C$ substrate temperature is illustrated in Fig. 5.2 (left). The fundamental transverse mode is lasing at 894.6 nm. A higher-order mode is located on the short-wavelength side with an SMSR of more than 40 dB, which well exceeds the target value of 20 dB stated in Sect. 2.3.4. Figure 5.2 (right) shows the spectrum of the VCSEL with $D_a = 4.6$ µm. At $T = 80°C$ and $I = 1.7$ mA, this device reaches the target wavelength with an SMSR of 33 dB. The experimental setup employed to measure the LIV characteristics and emission spectra is shown and described in App. F.1.

5.1.2 Polarization Control

Polarization stability of the VCSEL emission is of great interest for Cs-based atomic clock microsystems. Experimental characteristics of some inverted grating VCSELs are intro-

5.1 Static Characteristics

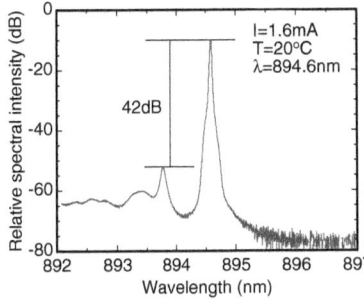
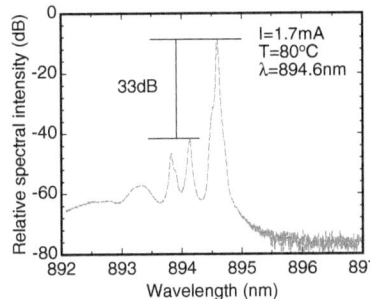

Figure 5.2: Emission spectra of the VCSELs from Fig. 5.1, where the device with $D_a = 3.5\,\mu m$ is operated at $I = 1.6\,mA$ and $T = 20°C$ (left) and the device with $D_a = 4.6\,\mu m$ is operated at $I = 1.7\,mA$ and $T = 80°C$ (right).

duced in what follows. Experimental results of other types of surface grating VCSELs, e.g., regular grating and inverted grating relief VCSELs are going to be introduced in Sect. 6.2. As mentioned earlier, the VCSELs incorporated in such microsystems experience high ambient temperatures, therefore, polarization stability has been investigated under elevated temperature conditions. As a representative example, Fig. 5.3 (left) shows polarization-resolved light–current–voltage (PR-LIV) characteristics of an inverted grating VCSEL with 4.6 µm active diameter, measured at $T = 65°C$. The dash-dotted and dashed lines indicate the optical powers P_{orth} and $5 \cdot P_{par}$ (for better clarity) measured behind a Glan–Thompson polarizer whose transmission direction is oriented orthogonal and parallel to the grating lines, respectively. The losses of the polarizer are responsible for the total power in Fig. 5.3 (left) being larger than $P_{orth} + P_{par}$. As expected from the simulation results shown in Sect. 4.4.3, the dominant polarization is orthogonal to the grating lines. The device remains polarization-stable from its threshold current of approximately 0.5 mA up to $I = 5\,mA$ with a maximum magnitude of OPSR as high as 21 dB. The OPSR is calculated from the ratio of the powers P_{par} and P_{orth} as

$$\text{OPSR} = 10 \cdot \log\left(\frac{P_{par}}{P_{orth}}\right) \tag{5.1}$$

and is displayed in Fig. 5.3 (left) with green symbols.

Figure 5.3 (right) depicts the polarization-resolved spectra at 65°C. The target wavelength is reached at $I = 2.2\,mA$ with an SMSR of nearly 50 dB and a peak-to-peak difference between the dominant and the suppressed polarization modes of 29 dB, which are much larger than the target values of 20 dB stated in Sect. 2.3.4. However, for the same current, the magnitude of the OPSR calculated from the powers in the two polarizations is only

5 Experimental Characterization of Atomic Clock VCSELs

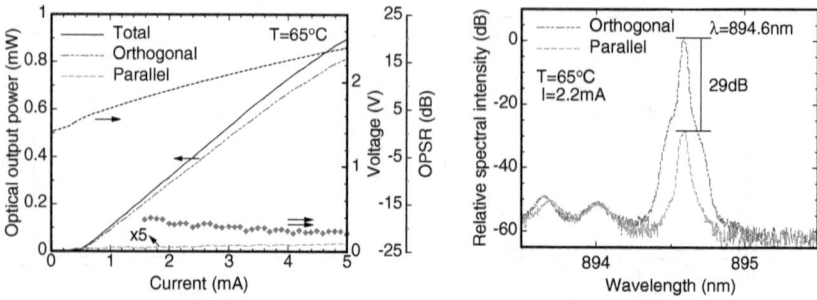

Figure 5.3: Polarization-resolved operation characteristics of an inverted grating VCSEL with $D_a = 4.6\,\mu m$ at $T = 65°C$ (left) and its polarization-resolved spectra at $I = 2.2\,mA$ (right). The surface grating has 70 nm depth and 0.6 µm period. The laser has $x = 4.5\%$ and $N_p = 28$.

19.2 dB. This difference originates from the integration of the spectral intensity over the complete sensitivity range of the photodiode used to measure the optical powers. Thus, all the power of the spontaneous emission and of the suppressed modes is included. Having $N_p = 28$, the laser from Fig. 5.3 shows a lower threshold current of approximately 0.5 mA due to increased reflectivity of its upper DBR compared to the device with the same D_a from Fig. 5.1 which has $N_p = 25$ and shows a threshold current of 0.65 mA.

Another inverted grating VCSEL emitting at 894.6 nm wavelength but at $T = 80°C$ is introduced in Fig. 5.4. The laser has 2.7 µm active diameter and shows approximately 0.5 mA threshold current, as depicted in Fig. 5.4 (left). The dash-dotted and dashed lines indicate the optical powers P_{orth} and $25 \cdot P_{par}$ (for better clarity), respectively. The dominant polarization is orthogonal to the grating lines and the device stays polarization-stable from its threshold up to $I = 4\,mA$ with a maximum magnitude of OPSR as high as 32 dB. Figure 5.4 (right) depicts the polarization-resolved high-temperature spectra. The target wavelength is reached at $I = 2.9\,mA$ with an SMSR of almost 30 dB and a peak-to-peak difference between the dominant and the suppressed polarization modes of 40 dB. Both the SMSR and the peak-to-peak spectral difference far exceed the target values of 20 dB.

The polarization control induced by the inverted grating has also been investigated for different substrate temperatures T. Figure 5.5 shows PR-LI characteristics of an inverted grating VCSEL with 4.4 µm active diameter, measured at T varied between 20 and 80°C in steps of 20°C. As can be seen, the VCSEL remains polarization-stable even well above thermal roll-over with a maximum magnitude of the OPSR as high as 20 dB. The threshold current increases with increasing T and it exceeds 1 mA at 80°C. Such increase in I_{th} is

5.1 Static Characteristics

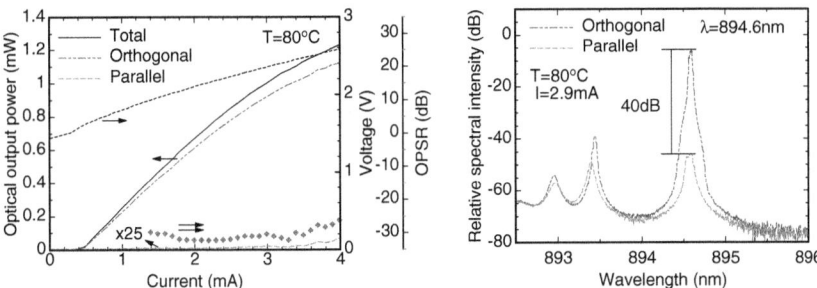

Figure 5.4: Polarization-resolved operation characteristics of an inverted grating VCSEL with $D_\mathrm{a} = 2.7\,\mathrm{\mu m}$ at $T = 80°C$ (left) and its polarization-resolved spectra at $I = 2.9\,\mathrm{mA}$ (right). The surface grating has 70 nm depth and 0.6 μm period. The laser has $x = 4\%$ and $N_\mathrm{p} = 25$.

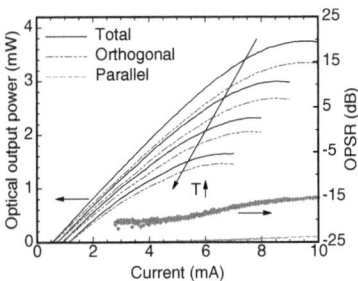

Figure 5.5: Polarization-resolved operation characteristics of an inverted grating VCSEL with $D_\mathrm{a} = 4.4\,\mathrm{\mu m}$, measured at $T = 20$ to $80°C$ in steps of $20°C$. The grating depth is 60 nm and the grating period is 0.7 μm. The laser has $x = 6\%$ and $N_\mathrm{p} = 25$.

consistent with the experimental results shown in Fig. 4.9 (left) for $x = 6\%$ indium content in the QWs.

5.1.3 Far-Field Properties

For optimum light routing in the atomic clock microsystem, it is important to know the beam properties of the VCSELs. From the simple grating theory, the far-field angle θ_SL, at which the maximum of the diffraction side-lobes with the integer order $m_\mathrm{d} \geq 1$ can be observed, is given by [200]

$$\theta_\mathrm{SL} = \sin^{-1}(m_\mathrm{d}\lambda/\Lambda)\,. \tag{5.2}$$

For $\Lambda < \lambda$ (i.e., sub-emission-wavelength grating periods), no side-lobes exist at all. Figure 5.6 (left) shows the emission far-fields orthogonal to the grating lines of two single-mode inverted grating VCSELs with grating periods of $\Lambda = 0.6$ and $0.7\,\mathrm{\mu m}$. There are no

87

5 Experimental Characterization of Atomic Clock VCSELs

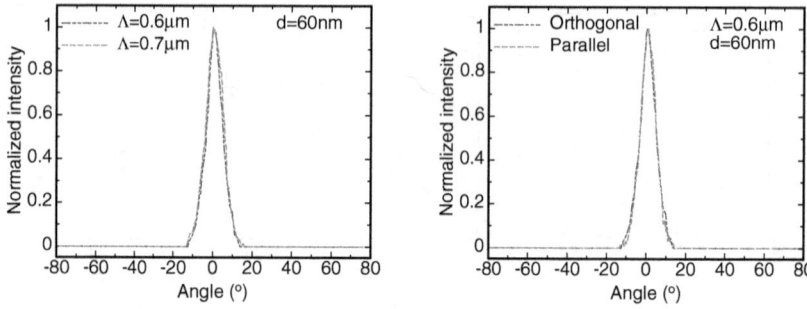

Figure 5.6: Emission far-fields measured orthogonal to the grating lines for two inverted grating VCSELs with 0.6 µm (dash-dotted) and 0.7 µm (dashed) grating periods (left), and emission far-fields measured orthogonal (dash-dotted) and parallel (dashed) to the grating lines for the VCSEL with 0.6 µm grating period (right). Both lasers have $D_a = 3.5$ µm, $x = 6\%$ and $N_p = 25$. The surface gratings have 60 nm grating depth. The measurements were done at $T = 20°C$ and $I = 4$ mA.

emission side-lobes, which proves the absence of diffraction effects by the sub-emission-wavelength grating. The emission far-fields of the device with $\Lambda = 0.6$ µm orthogonal and parallel to the grating lines are shown in Fig. 5.6 (right). Both far-fields are almost identical, indicating an almost perfectly circular beam shape. The experimental setup employed to measure the emission far-fields is shown and described in App. F.2.

The optical intensity in the main lobe of the VCSEL far-field can be most easily approximated by a Gaussian field profile as [66]

$$S_{FF}(\theta) = S_{FF}(0) \exp\{-2(\theta/\theta_0)^2\}, \quad (5.3)$$

where $S_{FF}(0)$ is a constant, θ is the divergence angle and $\theta_0 \approx \lambda/(\pi w_0)$[1]. w_0 is the field radius of the laser beam waist where its longitudinal position is assumed to be at the laser active layer[2]. From (5.3), the full far-field angle at which the intensity drops to one-half of its maximum is

$$\theta_{FWHM} = \sqrt{2\ln 2}\theta_0 \approx 1.18\lambda/(\pi w_0). \quad (5.4)$$

Figure 5.7 displays the emission far-field measured parallel to the grating lines of the VCSEL from Fig. 5.6 (right). The solid line is a Gaussian curve fit according to (5.3) with $\theta_0 = 7.9°$, $\theta_{FWHM} = 9.3°$ and $w_0 = 2$ µm. The diameter of the beam waist $2w_0 = 4$ µm

---
[1] The approximation is valid for small angles θ satisfying $\tan \theta \approx \theta$.
[2] Referring to Fig. 3.2, the longitudinal position of the beam waist could be assumed to be at the center of the second QW.

5.2 Dynamic Characteristics

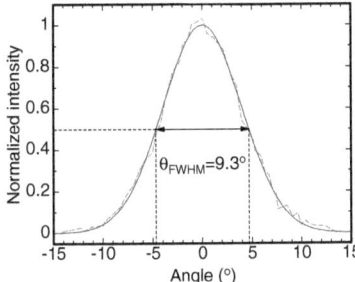

Figure 5.7: Emission far-field (dashed line) measured parallel to the grating lines of the VCSEL from Fig. 5.6 (right) along with a Gaussian fit profile (solid line) according to (5.3).

is slightly larger than the active diameter $D_a = 3.5\,\mu\text{m}$ of the VCSEL. This is expected due to the weak waveguiding typically present in such single-mode VCSELs with small D_a. By a curve fit of the emission far-field of the VCSEL with $0.7\,\mu\text{m}$ grating period from Fig. 5.6 (left), one obtains $\theta_0 = 8.9°$, $\theta_{\text{FWHM}} = 10.5°$ and $w_0 = 1.85\,\mu\text{m}$. The diameter of the beam waist is $2w_0 = 3.7\,\mu\text{m}$.

5.2 Dynamic Characteristics

Cs-based miniaturized atomic clocks require VCSELs with modulation bandwidths exceeding 5 GHz at low driving currents of, e.g., 2 mA. In this section, the dynamic characteristics of flip-chip-bondable VCSEL chips are presented with a main focus on the small-signal modulation response. The RIN and the microwave reflection spectra of the VCSEL chips are measured as approaches to determine their intrinsic modulation behavior.

5.2.1 Small-Signal Modulation Response

Small-signal modulation response curves of flip-chip-bondable VCSELs have been measured. In what follows, all modulation response functions relate the square of the fluctuations of the optical output power to those of the modulating electrical power. In other words, they relate the RF electrical power at the photodetector output to the RF electrical power which modulates the VCSEL. Therefore, it should be emphasized that the associated 3 dB corner frequency $f_{3\,\text{dB}}$ corresponds to an only 1.5 dB decay of the modulated optical signal. The experimental setup employed to measure the small-signal modulation response curves is shown and described in App. F.3. Small-signal modulation response curves of a VCSEL with 3.1 μm active diameter are depicted in Fig. 5.8 (left) for $T = 80°\text{C}$ at different bias currents. Figure 5.8 (right) displays the LIV characteristics of the VCSEL

5 Experimental Characterization of Atomic Clock VCSELs

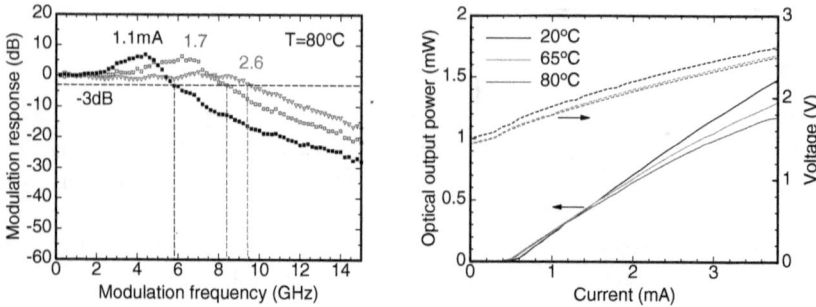

Figure 5.8: Small-signal modulation response curves of a VCSEL with $D_a = 3.1\,\mu\text{m}$ at different bias currents and $T = 80°C$ (left), and LIV characteristics of the same device at $T = 20, 65$ and $80°C$ (right). The laser has $x = 4\%$ and $N_p = 25$.

at $T = 20, 65$ and $80°C$. The threshold current is $0.54\,\text{mA}$ at $20°C$ and $0.45\,\text{mA}$ at $80°C$ with an intermediate minimum of $0.44\,\text{mA}$ at $65°C$. As usual it is observed that at higher bias currents, the resonance frequency f_r and the damping coefficient γ increase, thus the resonance peak becomes more damped and shifts to higher frequencies. For $T = 80°C$, the maximum $3\,\text{dB}$ bandwidth of about $9.5\,\text{GHz}$ is obtained at $I = 2.6\,\text{mA}$. A bandwidth of $5.8\,\text{GHz}$, already exceeding the target value of $5\,\text{GHz}$, is obtained at $I = 1.1\,\text{mA}$, i.e., only $0.65\,\text{mA}$ above threshold.

In order to determine the MCEF from (3.40), $3\,\text{dB}$ corner frequencies of the VCSEL are extracted from the small-signal modulation responses and plotted against $\sqrt{I - I_{\text{th}}}$ in Fig. 5.9. Equivalent measurements of modulation responses of the same VCSEL were also performed at $T = 20$ and $65°C$ and the $f_{3\,\text{dB}}$ values are included in Fig. 5.9. The MCEFs are determined from linear fits up to $\sqrt{I - I_{\text{th}}} = 1.3\,\sqrt{\text{mA}}$, above which thermal limitation[3] of $f_{3\,\text{dB}}$ starts to appear, particularly for $T = 65$ and $80°C$. MCEFs as high as $7.5\,\text{GHz}/\sqrt{\text{mA}}$ are achieved and do not decrease by more than 5% as T increases from 20 to $80°C$. The smaller MCEF at higher temperature is likely due to reductions of the current injection efficiency η_{I} and the differential gain coefficient $\bar{a}$.

5.2.2 Intrinsic Modulation Behavior

The small-signal modulation response curves in Fig. 5.8 (left) show the superposition of the intrinsic response of the VCSEL and of the electrical parasitic response according to (3.33). The modulation response of the electrical parasitics is attributed to the structure

---
[3]More explanations about thermal effects on laser bandwidth can be found at the end of Sect. 3.5.2.

5.2 Dynamic Characteristics

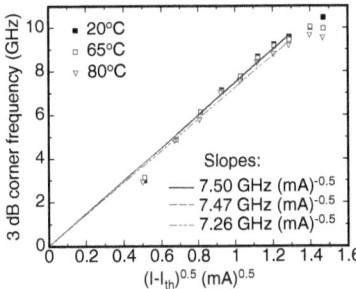

Figure 5.9: 3 dB corner frequency in dependence of $(I - I_{th})^{0.5}$ for the VCSEL from Fig. 5.8 at $T = 20$, 65 and 80°C. The MCEFs are the slopes of the linear fit lines.

and geometry of the VCSEL chip and can be represented by an electrical equivalent-circuit model. In order to define the intrinsic modulation behavior of a VCSEL, two additional RF measurement techniques are applied in this work. The first technique is based on measurements of the RIN spectra [94]. The second technique consists of measuring the microwave reflection spectra $S_{11}(f)$ from which the electrical parasitic components of the equivalent-circuit model can be extracted and its bandwidth limitation be calculated [201]. It should be emphasized that each method allows to determine the intrinsic modulation characteristics of the VCSEL. Both techniques are employed in order to improve the consistency of the results. In what follows these two techniques are presented in more detail.

Relative Intensity Noise

The experimental setup employed to measure the RIN spectra is shown and described in App. F.4. The RIN spectra for different currents and fit curves using (3.53) are depicted in Fig. 5.10 (left) for a VCSEL chip with 3.6 µm active diameter, measured at $T = 20°C$. Both f_r and γ have been extracted. From the fit parameters, the K-factor is then obtained as the slope of γ plotted against f_r^2, depicted in Fig. 5.10 (right). The resultant K-factor is 0.38 ns and the damping-limited maximum 3 dB corner frequency $f_{max,d}$ is 23.4 GHz according to (3.41). A drawback of this technique is its low dynamic range due to the high noise floor originating mainly from the thermal noise of the optical receiver and to smaller extent from the thermal noise of the low-noise amplifier (LNA). Shot noise plays a subordinate role. Therefore, increasing the dynamic range demands reducing the thermal noise and consequently smaller electrical bandwidths of the employed optical receiver and the LNA. The RIN measurements were found to be thermal noise-limited and not shot noise-limited. The (f_r^2, γ) data points from RIN measurements are available only in the low-frequency range due to the limited bandwidths of the employed LNA and the optical receiver (less than 10 GHz).

91

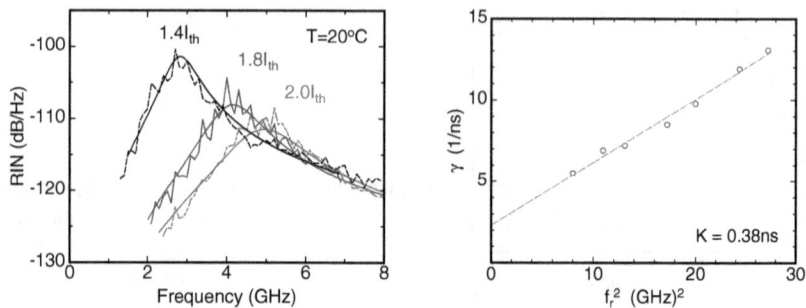

Figure 5.10: RIN spectra of a VCSEL chip with $D_a = 3.6\,\mu m$ at different drive currents and $T = 20°C$ (left). The solid lines are curve fits according to (3.52). Damping coefficient γ versus resonance frequency f_r squared obtained from the curve fits (right).

Microwave Reflection Spectrum

Figure 5.11 depicts the cross-section of a flip-chip-bondable atomic clock VCSEL together with an electrical equivalent-circuit model which accounts for the electrical parasitics attributed to the structure and geometry of the chip. The model is quite similar to the one proposed in Ref. [202] with additional inductance to account for the metal track. The model shows low-pass characteristics that affect the device performance by shunting the modulation current outside the active region at high modulation frequencies. The pad capacitance C_{pad} represents the capacitance between the metal pad on the polyimide passivation layer and the n-contact metal layer. The electrical resistances of the metal contacts and n- and p-type DBRs are included in the mirror resistance R_m. L and R_a represent the inductance of the metal track and the resistance associated with the oxide aperture, respectively. C_a is a combination of the capacitances of the diode junction of the active region C_j, of the oxide layer C_{ox}, and of the depletion layer formed under the oxide layer C_{dep} [203]. C_{ox} and C_{dep} are in series connection and both are parallel to C_j. Under forward bias, C_j is dominated by the diffusion capacitance C_{diff} which results from the diffusion of charge carriers through the active region [204]. C_a can be thus expressed as

$$C_a = C_{diff} + \left(\frac{1}{C_{ox}} + \frac{1}{C_{dep}}\right)^{-1}. \tag{5.5}$$

Above laser threshold, the active region is modeled as a short-circuit for small-signal conditions due to Fermi level pinning. At low frequencies, the VCSEL impedance is real and is given by the sum of R_m and R_a. As the frequency increases, C_a dominates over R_a,

5.2 Dynamic Characteristics

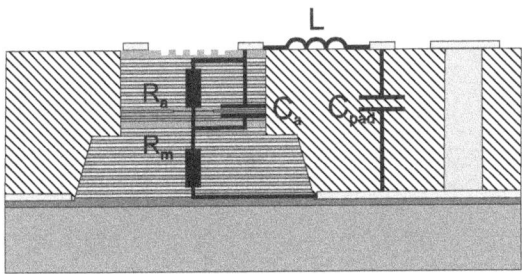

Figure 5.11: Schematic drawing of a flip-chip-bondable VCSEL structure including an electrical equivalent-circuit model.

and the real part of the impedance reduces to R_m. The input impedance of the electrical equivalent circuit in Fig. 5.11 can be written as

$$Z(f) = \left(\mathrm{i}2\pi f C_\mathrm{pad} + \left(\mathrm{i}2\pi f L + \frac{R_\mathrm{a}}{1 + \mathrm{i}2\pi f C_\mathrm{a} R_\mathrm{a}} + R_\mathrm{m} \right)^{-1} \right)^{-1}$$

$$= \frac{\tilde{A}_3(\tilde{A}_1 + \tilde{A}_2)}{\tilde{A}_1 + \tilde{A}_2 + \tilde{A}_3} \tag{5.6}$$

with

$$\tilde{A}_1 = \frac{R_\mathrm{a}}{1 + \mathrm{i}2\pi f C_\mathrm{a} R_\mathrm{a}}, \tag{5.7}$$

$$\tilde{A}_2 = \mathrm{i}2\pi f L + R_\mathrm{m} \tag{5.8}$$

and

$$\tilde{A}_3 = \frac{1}{\mathrm{i}2\pi f C_\mathrm{pad}}, \tag{5.9}$$

where $\mathrm{i} = \sqrt{-1}$ is the imaginary unit. The electrical reflection coefficient or the scattering (S-)parameter S_{11} resulting from the impedance $Z(f)$ is

$$S_{11} = \frac{Z(f) - Z_0}{Z(f) + Z_0}, \tag{5.10}$$

where Z_0 is the characteristic impedance of the measurement system, which is usually $50\,\Omega$. Therefore, the equivalent impedance of the VCSEL can be determined from the measured S_{11} using

$$Z(f) = Z_0 \frac{1 + S_{11}(f)}{1 - S_{11}(f)}. \tag{5.11}$$

5 Experimental Characterization of Atomic Clock VCSELs

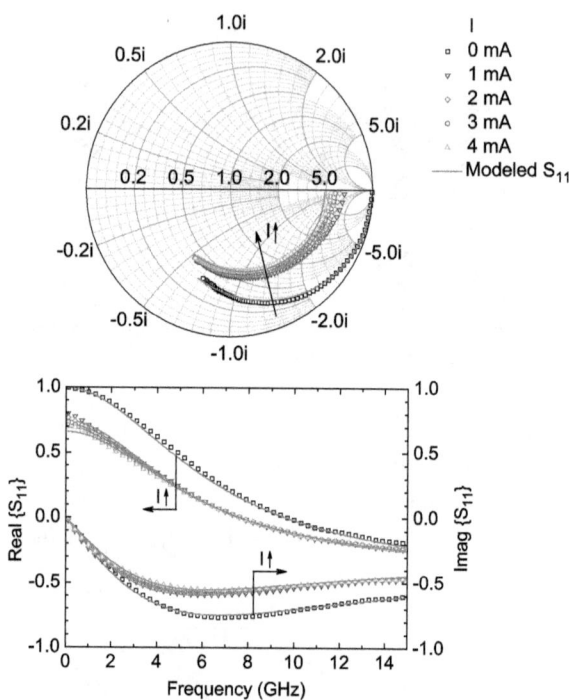

Figure 5.12: Measured S_{11} spectra of a VCSEL chip with 3.6 μm active diameter at $T = 20°C$ for an unbiased state as well as at 1, 2, 3 and 4 mA bias currents on a Smith chart over a frequency range from 0.1 to 15 GHz in 100 MHz steps (top) and their real and imaginary parts (bottom). The solid lines are fits to the measurement data and represent the modeled S_{11} using the equivalent circuit depicted in Fig. 5.11. The laser has $x = 6\%$ and $N_p = 25$.

In order to determine the intrinsic modulation behavior of the VCSEL chip, it is necessary to extract the electrical parasitic components of its equivalent-circuit model. The microwave reflection spectra $S_{11}(f)$ of VCSEL chips are measured at different bias currents I using a 50 Ω network analyzer[4] in order to get the associated input impedance spectra $Z(f)$ from (5.11). In this method, the equivalent-circuit parameters are varied until a best fit between calculated S_{11} from the equivalent-circuit elements and measured S_{11} is obtained.

Figure 5.12 (top) depicts the measured $S_{11}(f)$ spectra of a VCSEL that is nominally

---
[4]Spectrum analyzer HP8510C, S-parameter test-set HP8517A, and synthesized sweeper HP83651A.

5.2 Dynamic Characteristics

Table 5.1: Extracted values of the equivalent-circuit elements and their 3 dB corner frequencies at different I and $T = 20°C$ for the VCSEL chip from Fig. 5.12. The active diameter is 3.6 µm, the p-mesa diameter is 43 µm and the polyimide thickness is 7.5 µm.

I (mA)	L (pH)	R_m (Ω)	R_a (Ω)	C_pad (fF)	C_a (fF)	f_p (GHz)	$R_\mathrm{a} + R_\mathrm{m}$ (Ω)	R_s (Ω)
0	70	47	> 20 000	150	220	6.20	> 20 000	> 20 000
1	70	47	271	180	376	5.15	318	470
2	70	47	230	180	392	5.20	277	395
3	70	47	205	180	403	5.25	252	335
4	70	47	191	180	414	5.25	238	300

identical to the one from Fig. 5.10 on a Smith chart over frequencies f from 0.1 to 15 GHz in 100 MHz steps for an unbiased state and different biased states above threshold at $T = 20°C$. The VCSEL has an active diameter of 3.6 µm, a p-mesa diameter of 43 µm and a polyimide thickness of 7.5 µm. The real and imaginary parts of the measured S_{11} data are additionally plotted using Cartesian coordinates in Fig. 5.12 (bottom). Applying the electrical equivalent-circuit model in Fig. 5.11, the values for the parasitic elements shown in Table 5.1 could be extracted from curve fits which are plotted (solid curves) in Fig. 5.12 (top) and (bottom) using (5.10) and (5.6).

The resulting L, R_m, and C_pad do not show significant variations with I. The observed increase of C_a with increasing I is a consequence of the increase of C_diff [204] from (5.5). The resistance R_a strongly decreases from 271 Ω at 1 mA to 191 Ω at 4 mA. This behavior is consistent with the decrease of the differential series resistance R_s — extracted by linear interpolation of the IV curve — at high operating currents. However, R_s is larger than the sum of R_m and R_a, as stated in Table 5.1. This difference is due to a small misfit in the low-frequency range ($f < 1$ GHz) between the measured and the modeled Real$\{S_{11}\}$, as can be seen in Fig. 5.12 (bottom). The non-ideal fit in the low-frequency range is a compromise to achieve an almost perfect curve fit in the high-frequency range ($f > 1$ GHz). The relatively high value of > 20 kΩ for the unbiased state points out the open-circuit characteristic of the VCSEL. From the extracted values of the circuit elements, the electrical parasitic 3 dB corner frequency f_p can be determined for different I from the squared magnitude $|M_\mathrm{p}(f)|^2$ of the parasitic transfer function

$$M_\mathrm{p}(f) = \frac{V_{R_\mathrm{a}}}{V_\mathrm{s}} = \frac{\tilde{A}_1 \tilde{A}_3}{\tilde{A}_3(\tilde{A}_1 + \tilde{A}_2) + Z_0(\tilde{A}_1 + \tilde{A}_2 + \tilde{A}_3)}, \quad (5.12)$$

where V_s and V_{R_a} are the small-signal modulating voltages generated by the source and reaching the active region of the VCSEL represented by R_a and C_a in Fig. 5.11, respectively.

5 Experimental Characterization of Atomic Clock VCSELs

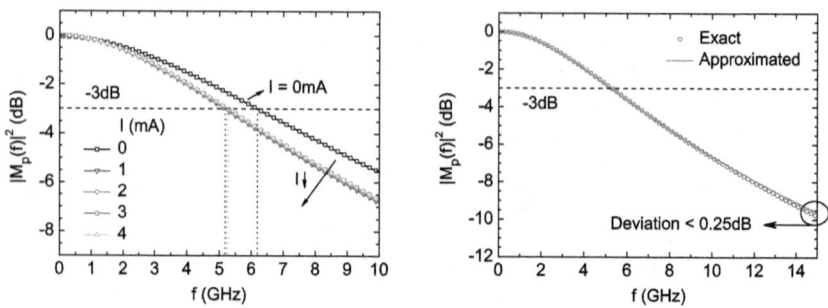

Figure 5.13: Calculated parasitic transfer function $|M_\mathrm{p}(f)|^2$ in dB of the VCSEL from Fig. 5.12 for an unbiased state as well as at 1, 2, 3 and 4 mA bias currents using (5.12) by inserting the values of the extracted parasitic elements shown in Table 5.1 (left). Exact $|M_\mathrm{p}(f)|^2$ (circles) together with its approximation (solid line) for $I = 3\,\mathrm{mA}$ using (5.12) and the second term in (3.33), respectively (right). A deviation of less than 0.25 dB between exact and approximated $|M_\mathrm{p}(f)|^2$ at $f = 15\,\mathrm{GHz}$ is indicated by a small ellipse.

Figure 5.13 (left) depicts the parasitic transfer function $|M_\mathrm{p}(f)|^2$ in dB versus the modulation frequency f of the VCSEL from Fig. 5.12 for the same operation modes. The corner frequencies f_p are indicated as vertical lines for different bias currents in Fig. 5.13 (left) and added to Table 5.1. The current-dependent behavior of R_a and C_a (see Table 5.1) results in a very slight increase of f_p with I until reaching a maximum of 5.25 GHz at $I = 3$ or 4 mA. It is worth to note that $|M_\mathrm{p}(f)|^2$ from (5.12) can be very well approximated by a first-order low-pass term as the one used in (3.33), resulting in a deviation of less than 0.25 dB up to a frequency of 15 GHz, as can be seen in Fig. 5.13 (right).

Using the modulation transfer function (3.33) and the extracted values of f_p, precise fits of the measured modulation response curves can be performed and f_r and γ can be extracted. Figure 5.14 (left) shows the fit curves superimposed on the experimental data for the VCSEL from Fig. 5.12.

The (f_r^2, γ) data points are plotted in Fig. 5.15 (as square symbols), along with the data points extracted from the RIN fits of Fig. 5.10 (as circles). The resultant K-factor is 0.36 ns and the damping-limited maximum 3 dB corner frequency $f_\mathrm{max,d}$ is 24.7 GHz according to (3.41), with a high level of consistency between data points from the small-signal response and the RIN approach. However, the extraction procedure was more difficult from the RIN curves compared to the modulation response curves, especially for the damping coefficients which depend somewhat on the initial estimates. In order to obtain the MCEF and D-factor, both f_3dB and f_r are extracted from Fig. 5.14 (left) and plotted

5.2 Dynamic Characteristics

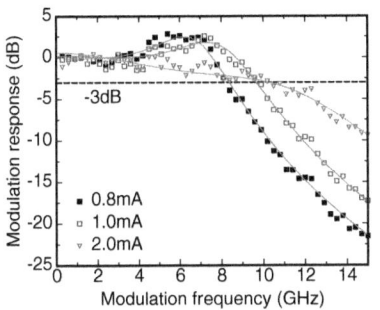
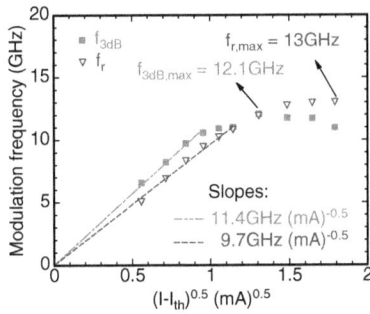

Figure 5.14: Small-signal modulation response curves for the VCSEL from Fig. 5.12 at different bias currents and $T = 20°C$ (left). The solid lines are curve fits according to (3.33). 3 dB corner frequency $f_{3\,dB}$ (square symbol) and resonance frequency f_r (triangle symbol) versus $(I - I_{th})^{0.5}$ (right). The two lines (right) are linear fits, where their slopes are given and represent the MCEF (dash-dotted) and the D-factor (dashed), respectively. The maximum 3 dB corner frequency $f_{3\,dB,max}$ and the maximum resonance frequency $f_{r,max}$ are indicated.

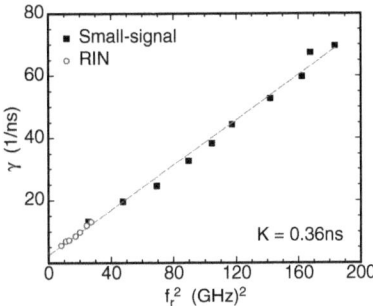

Figure 5.15: Damping coefficient γ versus resonance frequency f_r squared of a VCSEL with 3.6 µm active diameter. Square symbols are the values obtained by curve fitting of measured small-signal modulation characteristics shown in Fig. 5.14 (left), while circles are the values obtained by fitting the measured RIN spectra shown in Fig. 5.10.

against $\sqrt{I - I_{th}}$ in Fig. 5.14 (right). The resultant MCEF and D-factor are 11.4 and 9.7 GHz/$\sqrt{mA}$, respectively. At small bias currents, $f_{3\,dB}$ and f_r increase linearly with $\sqrt{I - I_{th}}$ to 10.6 and 10.9 GHz at $I = 1.2$ and 1.6 mA, respectively, then they deviate from linearity at larger I and reach maximum frequencies of $f_{3\,dB,max} = 12.1$ GHz and $f_{r,max} = 13$ GHz, respectively.

5 Experimental Characterization of Atomic Clock VCSELs

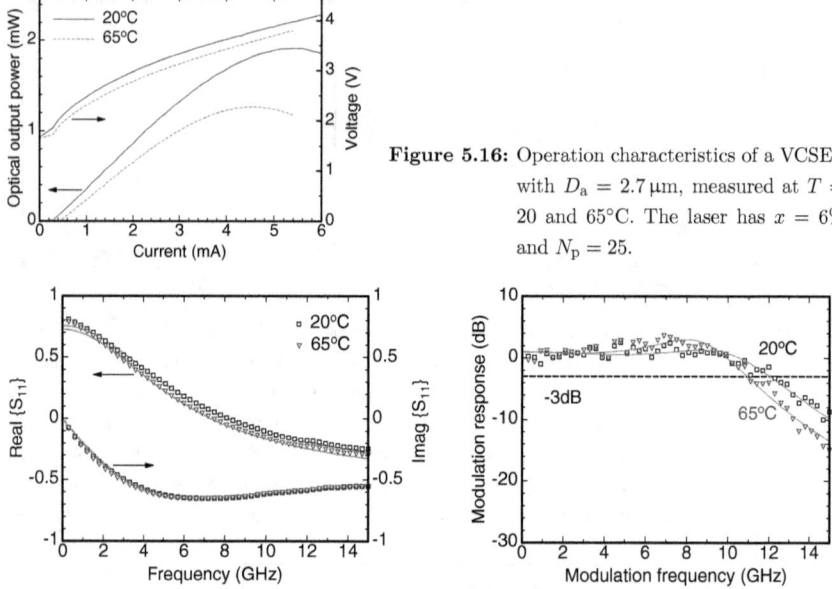

Figure 5.16: Operation characteristics of a VCSEL with $D_a = 2.7\,\mu\text{m}$, measured at $T = 20$ and $65\,°\text{C}$. The laser has $x = 6\%$ and $N_p = 25$.

Figure 5.17: Real and imaginary parts of measured $S_{11}(f)$ spectra (over a 15 GHz frequency span) of the VCSEL from Fig. 5.16 at 20 and 65°C and a bias current of $I - I_{\text{th}} = 1.15\,\text{mA}$ (left). The solid lines are curve fits using the equivalent-circuit model of Fig. 5.11. Small-signal modulation response curves of the VCSEL under the same temperatures and bias conditions (right). The solid lines are curve fits according to (3.33).

The resultant thermally-limited maximum 3 dB corner frequency $f_{\text{max,t}}$ is 20.2 GHz according to (3.42), where damping and electrical parasitic effects are neglected. The deviation from linearity is likely due to reductions of η_i and $\bar{a}$ (see (3.34)) as a result of device self-heating. Additionally to thermal effects, f_{3dB} exhibits earlier deviation from linearity at smaller bias due to the bandwidth limitation by electrical parasitics quantified by $f_p \approx 5.2\,\text{GHz}$ (see Table 5.1). Consequently, a comparison between $f_p \approx 5.2\,\text{GHz}$, $f_{\text{max,t}} = 20.2\,\text{GHz}$ and $f_{\text{max,d}} = 24.7\,\text{GHz}$ demonstrates that the electrical parasitic effects represent the major limitation to the device bandwidth. The intrinsic modulation behavior is determined at different substrate temperatures T for another VCSEL with a smaller active diameter of 2.7 µm, a slightly smaller p-mesa diameter of 42 µm and the same polyimide thickness as the VCSEL from Fig. 5.12.

Figure 5.16 displays the LIV characteristics of the VCSEL at 20 and 65°C. The threshold

5.2 Dynamic Characteristics

Table 5.2: Equivalent-circuit elements like in Table 5.1 for the VCSEL chip from Fig. 5.16 at $T = 20°C$. The active diameter is $2.7\,\mu m$, the p-mesa diameter is $42\,\mu m$ and the polyimide thickness is $7.5\,\mu m$.

I (mA)	L (pH)	R_m (Ω)	R_a (Ω)	C_pad (fF)	C_a (fF)	f_p (GHz)
0.5	70	45	434.5	187	284	5.9
1.0	70	45	348.0	187	287	6.1
2.0	70	45	271.0	187	301	6.2
3.0	70	45	225.0	187	313	6.3

current is 0.27 and $0.44\,\mathrm{mA}$, respectively. The increase of I_th with T is consistent with the experimental results illustrated in Fig. 4.9 (left) for $x = 6\%$ indium content in the QWs. The microwave reflection spectra $S_{11}(f)$ of the VCSEL have been measured with a $50\,\Omega$ network analyzer at different currents for both temperatures in order to obtain the associated input impedance spectra $Z(f)$. As an example, Fig. 5.17 (left) depicts the real and imaginary parts of the measured S_{11} spectra for the VCSEL chip at $T = 20$ and $65°C$ and a bias current of $I = I_\mathrm{th} + 1.15\,\mathrm{mA}$ along with curve fits from the equivalent-circuit model using (5.10) and (5.6). The extracted values of the equivalent-circuit elements are shown in Table 5.2 for different I and at $T = 20°C$. Similar to the VCSEL from Table 5.1, the resulting L, R_m and C_pad do not exhibit significant changes with I and their values are almost identical for both VCSELs. Moreover, the current-dependent behavior of R_a and C_a is similar. From the squared magnitude of the parasitic transfer function (5.12), f_p is determined for different I and added to Table 5.2. The extraction of parasitics were also performed for $T = 65°C$ and different I, resulting in slightly smaller f_p, particularly by $0.2\,\mathrm{GHz}$. Using the modulation transfer function (3.33) and the extracted values of f_p, precise fits of the measured modulation response curves can be performed and f_r and γ can be extracted. Fitted modulation response curves for the laser from Fig. 5.16 at $20°C$ and $65°C$ substrate temperatures and a bias current of $I = I_\mathrm{th} + 1.15\,\mathrm{mA}$ are displayed in Fig. 5.17 (right). It is noticed that the resonance peak shifts to lower frequency and the damping decreases with increasing substrate temperature even at equal $I - I_\mathrm{th}$ bias conditions, which can be attributed to a decrease of η_l and $\bar{a}$. In particular, f_r decreases from $10.5\,\mathrm{GHz}$ to $9\,\mathrm{GHz}$ and γ decreases from $36.9\,\mathrm{ns}^{-1}$ to $26.8\,\mathrm{ns}^{-1}$ for T being changed from 20 to $65°C$. This corresponds to a decrease of f_r and γ by factors of 0.85 and 0.72, respectively, where the latter is equal to the former squared, as could be expected from (3.38). According to (3.34), $f_\mathrm{r} \propto \sqrt{\eta_\mathrm{l} \cdot \bar{a}}$. Therefore, one can conclude that a decrease of $\eta_\mathrm{l} \cdot \bar{a}$ by a factor of 0.72 occurred for T being varied from 20 to $65°C$.

The K-factor is then obtained from the slope of γ plotted against f_r^2, as shown in Fig. 5.18.

5 Experimental Characterization of Atomic Clock VCSELs

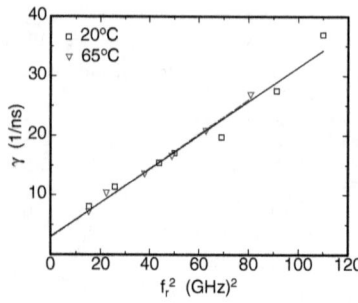

Figure 5.18: γ versus f_r^2 of the VCSEL from Fig. 5.17 along with linear fits at 20°C (solid) and 65°C (dashed).

It has a value of about 0.28 ns for both substrate temperatures. The damping-limited maximum 3 dB corner frequency $f_{\text{max,d}}$ is 31.5 GHz.

Chapter 6

Improved and Alternative Atomic Clock VCSELs

In this chapter, improvements and alternatives of atomic clock VCSELs are presented. First, the influence of modifying the number of top Bragg mirrors on the static characteristics of the VCSELs, such as threshold gain, threshold current, differential quantum efficiency and optical output power, are shown. For polarization control, so far only inverted grating VCSELs were experimentally introduced. In this chapter, experimental characteristics of alternative surface grating VCSELs will be presented along with their advantages and drawbacks compared to the inverted grating devices. Such techniques include regular gratings and inverted grating reliefs which were explained and discussed in Chap. 4. Two revised VCSEL chip designs will be introduced in this chapter. As a first approach, VCSELs with quite thinner polyimide and smaller mesa size, compared to the devices introduced in Chap. 5, have been fabricated. Such VCSELs show better electrical parasitic bandwidths. Later, the structure has been modified by further reducing both the polyimide thickness and the mesa size, which gives almost similar parasitic behavior and the benefit of reduced processing complexity compared to the first revised design. At the end of the chapter, preliminary reliability tests of some atomic clock VCSELs are reported.

6.1 Modification of the Top Bragg Mirrors

Etching a semiconducting surface grating in the top DBR of the VCSEL structure introduces optical losses by decreasing the reflectivity of the mirror. This increases the threshold gain and, hence, the threshold current. Also the differential quantum efficiency increases. This can be noticed in Fig. 6.1 (left), which depicts the LI characteristics of two VCSELs with and without an inverted grating at $T = 20°C$. Both VCSELs have nominally

6 Improved and Alternative Atomic Clock VCSELs

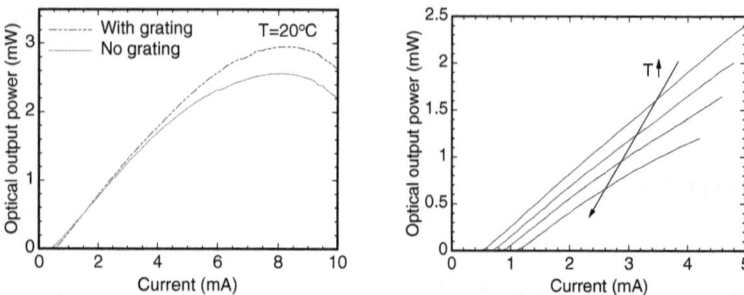

Figure 6.1: Operation characteristics of two VCSELs with and without inverted grating having nominally identical active diameters $D_a = 4.1\,\mu m$ (left) and an inverted grating VCSEL with $D_a = 4.4\,\mu m$, measured at $T = 20$ to $80°C$ in steps of $20°C$ (right). For the grating VCSELs, the surface grating has $60\,nm$ grating depth, $0.7\,\mu m$ grating period, and $50\,\%$ duty cycle. All VCSELs have $x = 6\%$ indium content in the QWs and $N_p = 25$ top mirror pairs.

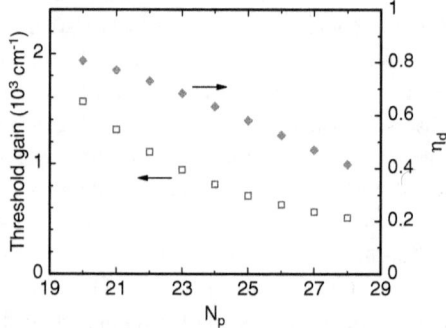

Figure 6.2: Calculated threshold gains and differential quantum efficiencies of the VCSELs fabricated during the research performed for this dissertation, where the number of top mirror pairs is varied.

identical active diameters. The threshold current of the inverted grating VCSEL is still less than $1\,mA$. However, the VCSELs to be incorporated in the MAC microsystems will experience high ambient temperatures (e.g., $T = 65$ to $80°C$). As can be seen in Fig. 6.1 (right), the threshold current increases as the substrate temperature increases until it exceeds[1] $1\,mA$ at $80°C$. The increase of threshold with increasing temperature can be mainly attributed to the detuning of the cavity resonance and the peak material gain

---

[1] The MAC-TFC threshold current limit is $1\,mA$, see Sect. 2.3.4.

6.1 Modification of the Top Bragg Mirrors

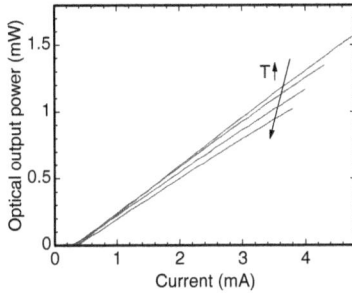
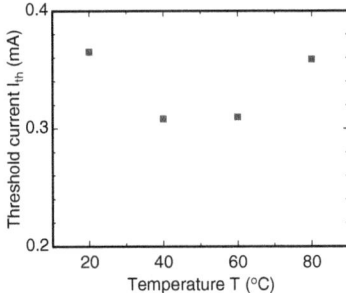

Figure 6.3: Operation characteristics at $T = 20$ to $80\,°\text{C}$ in steps of $20\,°\text{C}$ of an inverted grating VCSEL with $D_\text{a} = 5.1\,\mu\text{m}$, $x = 4.5\%$ and $N_\text{p} = 28$ (left). Threshold current dependence on temperature (right). The grating has 70 nm grating depth and 0.6 µm grating period.

and the decrease of the gain itself. This demands a higher carrier density to reach the threshold gain. In order to reduce the threshold current, one may increase the number of top mirror pairs N_p. Figure 6.2 depicts calculated threshold gains and differential quantum efficiencies η_d in dependence of N_p for the VCSEL layer structure utilized in this dissertation and depicted in App. E. The calculations were done using the one-dimensional transfer-matrix method [66]. The thickness of the cap layer t_cap is set to be $\lambda_\text{mat}/2$ (i.e., regular VCSEL structure with no surface grating). One can conclude that by adding, for instance, three mirror pairs to the VCSELs shown in Fig. 6.1, the threshold gain could be reduced by approximately 30 %. A similar trend can be observed for grating VCSELs from the three-dimensional simulation results depicted in Fig. 4.17 (left), where N_p is varied between 25 and 31 in steps of 3.

As depicted in Fig. 6.2, η_d (and accordingly P, see (3.16)) decreases with increasing N_p. For instance, with $N_\text{p} = 28$, η_d decreases by 30 % in comparison with the case of $N_\text{p} = 25$. Figure 6.3 (left) depicts LI characteristics of an inverted grating VCSEL with T varied between 20 and $80\,°\text{C}$ in steps of $20\,°\text{C}$. It can be noticed that the threshold current does not exceed 0.5 mA even at $80\,°\text{C}$. Compared to the VCSEL of Fig. 6.1 (right), this device has three additional top mirror pairs and 4.5% indium content in the QWs. These two modifications are the cause of the reduction in threshold current as well as in optical output power. As explained in Sect. 4.2.6, an optimized QW composition (e.g., indium content $x = 4.5\%$) contributes to improved device performance in terms of reduced threshold currents at elevated substrate temperatures (e.g., $T \approx 55\,°\text{C}$) as shown in Fig. 4.9 (left). This can be seen in Fig. 6.3 (right) by sub-mA threshold currents, ranging between 0.3 and 0.36 mA. An intermediate minimum at $T \approx 50\,°\text{C}$ is expected.

103

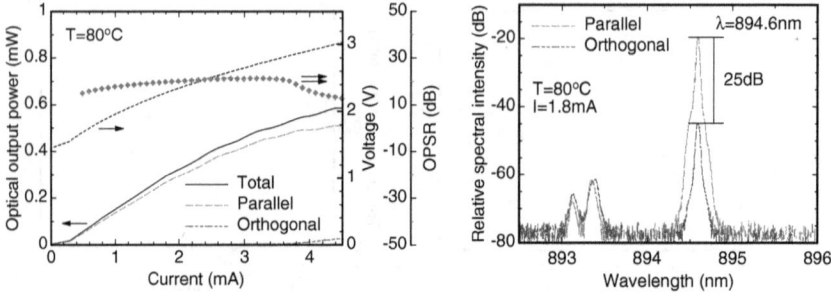

Figure 6.4: Polarization-resolved operation characteristics of a regular grating VCSEL with $D_\mathrm{a} = 3\,\mathrm{\mu m}$, measured at $T = 80°\mathrm{C}$ (left) and its polarization-resolved spectra at $I = 1.8\,\mathrm{mA}$ (right). The grating has $0.6\,\mathrm{\mu m}$ period and $120\,\mathrm{nm}$ depth. The laser has $x = 4\%$ and $N_\mathrm{p} = 25$.

6.2 Alternative Surface Grating Approaches

So far, the introduced atomic clock VCSELs employed inverted surface gratings, where the grating is etched in an extra topmost GaAs quarter-wave anti-phase layer. Experimental characteristics of such devices have been presented in Sect. 5.1.2. An alternative approach to full-area gratings is to etch a surface grating over a circular area of only 3 to 4 µm diameter in the center of the outcoupling facet. Such inverted grating relief VCSELs simultaneously provide favorable single-mode and single-polarization emission, as explained in Sect. 4.4.2. Owing to their larger active diameters, they exhibit the favorable property of reduced ohmic resistance and can thus provide greater potential for increased lifetime. However, inherently higher optical losses, even of the fundamental mode, and the larger active diameter itself lead to higher threshold currents compared to full-area grating VCSELs. Such true single-mode inverted grating relief VCSELs emitting at 894.6 nm wavelength have been fabricated for use in Cs-based MACs. Their experimental characteristics are presented in what follows. Another favorable alternative approach are the regular grating VCSELs explained in Sect. 4.4.2. Detailed numerical simulations for regular and inverted VCSEL designs were presented in Sect. 4.4.3 and showed that regular grating VCSELs have inherently lower threshold gains than equivalent inverted grating devices. Reduced threshold currents of regular grating VCSELs are thus expected. This functional property is highly desired for low-power atomic clock applications. Such low threshold, single-mode, and polarization-stable VCSELs emitting at 894.6 nm wavelength have been fabricated for use in Cs-based MACs. Their experimental characteristics are presented in what follows.

6.2 Alternative Surface Grating Approaches

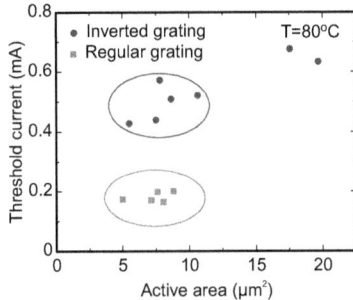

Figure 6.5: Experimentally obtained threshold currents of seven inverted grating VCSELs and five regular grating VCSELs with different active areas. All devices are single-mode. Inverted and regular grating devices with a similar range of active areas are grouped in two ellipses. $x = 4\%$, $N_p = 25$ and $T = 80°C$ for all VCSELs.

6.2.1 Regular Grating VCSELs

Figure 6.4 (left) depicts the PR-LIV characteristics of a regular grating VCSEL with an active diameter of 3 µm, a grating period of 0.6 µm and a grating depth of 120 nm. The optical powers of the two polarization modes are measured behind a Glan–Thompson polarizer by orienting its transmission direction parallel and orthogonal to the grating lines. The corresponding powers are denoted as P_par and P_orth. They are indicated in the figure by dashed and dash-dotted lines, respectively. The VCSEL remains polarization-stable from its threshold current of approximately 0.2 mA up to thermal roll-over with an average OPSR of 18.5 dB and a peak value of 21 dB. The average OPSR is calculated from the data for currents in steps of 0.1 mA and output powers corresponding to 10...100% of the maximum. Figure 6.4 (right) shows the polarization-resolved high-temperature spectra. The target wavelength of 894.6 nm is reached at a bias current $I = 1.8$ mA with an SMSR of 42 dB. The peak-to-peak difference between the dominant and the suppressed polarization modes is about 25 dB. Both the SMSR and the peak-to-peak difference between the dominant and suppressed polarization modes far exceed the target values of 20 dB. The dominant polarizations of the regular grating VCSELs are always found to be parallel to the grating lines, which is consistent with the simulation results depicted in Fig. 4.16 (right) for about 120 nm grating depth.

As it was shown by the simulation results in Sect. 4.4.3, for regular grating VCSELs the material threshold gain of the selected polarization mode is minimum at 120 to 130 nm grating depth d, which corresponds to about $\lambda_\mathrm{mat}/2$. The minimum threshold gain is approximately 30% of the corresponding minimum of the inverted grating VCSEL at $d = 60$

105

6 Improved and Alternative Atomic Clock VCSELs

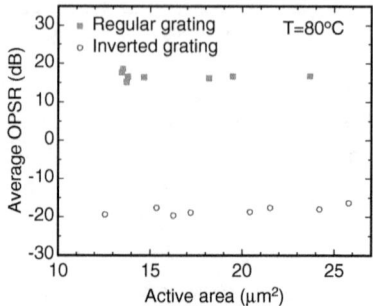

Figure 6.6: Experimentally obtained average OPSRs at $T = 80°C$ of several regular and inverted grating VCSELs of different active areas. The gratings have 0.6 µm period and 120 nm (70 nm) depth for regular (inverted) grating VCSELs. $x = 4\%$ and $N_p = 25$ for all devices.

to 70 nm, as can be noticed from Fig. 4.16. Reduced threshold currents of regular grating VCSELs are thus expected. Experimentally obtained I_{th} of several inverted grating and regular grating VCSELs having different active areas are displayed in Fig. 6.5. All the lasers are single-mode and have 0.6 µm grating period and 50% duty cycle, however, $d = 120$ nm and 70 nm for regular and inverted grating VCSELs, respectively. Regular grating VCSELs in the lower ellipse in Fig. 6.5 show an average I_{th} of about 0.2 mA, which is approximately 40% of the I_{th} of the inverted grating VCSELs in the upper ellipse within the same range of active areas between 5 and 10 µm².

Figure 6.6 illustrates experimentally obtained average OPSRs of several regular and inverted grating VCSELs with identical grating parameters of Fig. 6.5. Positive OPSRs indicate the dominant polarization modes of the regular VCSELs to be parallel to the grating lines. In contrast, the dominant polarization of the inverted grating VCSELs is orthogonal to the grating, thus OPSR < 0 dB. The regular grating devices from Fig. 6.6 show an average |OPSR| of about 16.8 dB which is approximately 92% of the average |OPSR| of the inverted grating VCSELs from the same figure. Therefore, the inverted grating lasers slightly outperform the regular grating devices in terms of polarization control. However, the regular grating devices provide much lower threshold operation, as it was shown in Fig. 6.5. The tendency of regular grating devices (having relatively large grating depths, e.g., $d = 120$ nm) to show lower |OPSR| was observed before [200]. It was pointed out (by similar experiments and simulations to those performed in this dissertation but at different wavelength) that surface gratings increase the contribution of suppressed polarization modes with increasing d, with which also the relative dichroism

6.2 Alternative Surface Grating Approaches

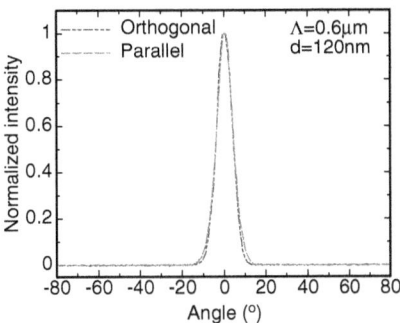

Figure 6.7: Emission far-fields of a regular grating VCSEL with $D_a = 4.8\,\mu m$, measured at $T = 20°C$. The almost overlapping dash-dotted and dashed lines indicate the normalized optical intensity orthogonal and parallel to the grating lines, respectively. The laser has $x = 4\%$ and $N_p = 25$.

increases. Considering the simulation results shown in Fig. 4.16, one can thus design regular grating lasers[2] with shallower etch depths (e.g., $d = 70\,nm$) which can still provide reduced threshold gains (e.g., by a factor of 0.7 at $d = 70\,nm$) compared to the inverted grating devices with the same d, and hence decreased threshold currents. Such shallower regular grating devices are expected to provide higher |OPSR| values which are comparable to those of the inverted grating VCSELs from Fig. 6.6. It is worth noting that VCSELs in Fig. 6.6 with active areas larger than about $20\,\mu m^2$ tend to be transverse multi-mode at higher currents. As is well known [12,174], surface gratings also stabilize the polarization of higher-order modes.

The effect of the surface grating on the beam properties of regular grating VCSELs is investigated by measuring the emission far-fields. Figure 6.7 shows the far-fields parallel and orthogonal to the grating lines of a regular grating VCSEL with $4.8\,\mu m$ active diameter at $I = 2\,mA$ and $T = 20°C$, where $I_{th} = 0.9\,mA$. The grating depth is $120\,nm$ and the period is $0.6\,\mu m$. The absence of side-lobes in the emission far-fields proves that VCSELs with sub-emission-wavelength grating periods do not suffer from diffraction losses in air. Almost identical far-field patterns along the parallel and orthogonal directions indicate a circular beam profile. The Gaussian-like shape reflects the single-mode emission of the VCSEL at the given current. Figure 6.8 displays the emission far-field orthogonal to the grating lines of the VCSEL from Fig. 6.7 along with a Gaussian curve fit (solid line) ac-

---
[2]The regular grating technique has been applied towards the end of the research performed for this dissertation. Therefore, optimized fabrication runs of regular grating VCSELs, unfortunately, could not be made.

107

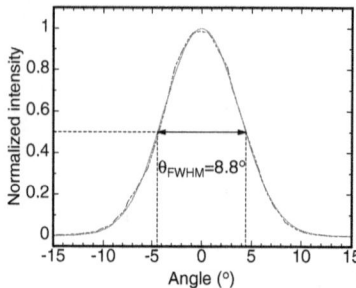

Figure 6.8: Emission far-field (dashed line) measured orthogonal to the grating lines of the VCSEL from Fig. 6.7 along with a Gaussian fit profile (solid line) according to (5.3).

cording to (5.3) with $\theta_0 = 7.5°$, $\theta_{\text{FWHM}} = 8.8°$ and $w_0 = 2.2\,\mu\text{m}$. Unlike the result shown in Fig. 5.7, the diameter of the beam waist $2w_0 = 4.4\,\mu\text{m}$ is slightly smaller than the active diameter $D_\text{a} = 4.8\,\mu\text{m}$ of the VCSEL from Fig. 6.8. This can be attributed to stronger waveguiding in the VCSEL with the larger D_a.

6.2.2 Inverted Grating Relief VCSELs

Figure 6.9 (a) displays an optical microscope image of a flip-chip-bondable VCSEL with an inverted grating relief. The VCSEL has $300 \times 300\,\mu\text{m}^2$ size. Figures 6.9 (b) and (c) show the grating relief and its surface profile measured with an AFM. Grating reliefs with 3 μm diameter, almost quarter-wave etch depth, sub-emission-wavelength grating periods of 0.6 μm and 50% duty cycle have been employed. The cross-section of the VCSEL is depicted in Fig. 4.22.

The PR-LIV characteristics of a grating relief VCSEL with 4.5 μm active diameter, measured at $T = 80°\text{C}$ substrate temperature are shown in Fig. 6.10 (left). The device shows a threshold current of approximately 0.8 mA which is higher compared to the full-area inverted and regular grating VCSELs from Fig. 6.5. Such an increased threshold current is caused by the inherently higher optical outcoupling losses, even of the fundamental mode, induced by the unetched anti-phase layer outside and inside the grating relief. The dash-dotted and dashed lines indicate the optical powers P_{orth} and P_{par}, respectively. The VCSEL remains polarization-stable from its threshold current up to thermal roll-over with an average OPSR of $-18.4\,\text{dB}$ and a maximum |OPSR| as high as 19.4 dB. The average OPSR is calculated from the data for currents in steps of 0.1 mA and output powers corresponding to 10 ... 100% of the maximum. Above $I \approx 5\,\text{mA}$, the dominant polarization mode starts to decrease in power due to thermal roll-over, while the other polarization mode starts to appear. This causes |OPSR| to decrease.

Figure 6.10 (right) depicts polarization-resolved spectra at 80°C. The target wavelength

6.2 Alternative Surface Grating Approaches

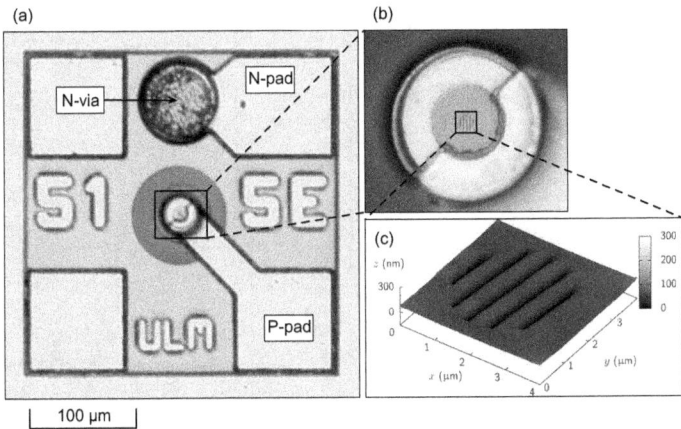

Figure 6.9: Optical micrograph of a fully processed flip-chip-bondable VCSEL with an inverted grating relief (a). Zoomed outcoupling facet (b). Surface profile within the grating relief region measured with an AFM (c). The grating relief has a diameter of 3 µm, a grating period of 0.6 µm and an etch depth of 70 nm.

of 894.6 nm is reached at a current of 3.8 mA with both an SMSR and a peak-to-peak difference between the dominant and the suppressed polarization modes of almost 27 dB, which well exceed the target values of 20 dB. However, for the same current, the magnitude of the OPSR calculated from the powers in the two polarizations is only 19.4 dB. As explained in Sect. 5.1, such a difference is due to the integration of the spectral intensity over the complete sensitivity range of the photodiode used to measure the optical powers. Therefore, the power of the spontaneous emission and of the suppressed modes is included. The dominant polarization of the inverted grating relief VCSELs is always found to be orthogonal to the grating lines, which is consistent with the simulation results depicted in Fig. 4.16 (left) for about 70 nm grating depth. The polarization control induced by the grating relief has also been investigated for different ambient temperatures T. Figure 6.11 depicts PR-LIV characteristics of a grating relief VCSEL with 5 µm active diameter, measured at T varied between 20 and 100°C in steps of 20°C. As can be seen, the VCSEL remains polarization-stable even well above thermal roll-over. The magnitudes of the OPSR for $T = 80$ and 100°C are increased in comparison with lower temperatures as the current exceeds 4.5 mA.

To investigate the enhancement of fundamental-mode emission, standard reference devices were fabricated on the same wafer adjacent to the grating relief VCSELs for comparison. For the reference VCSELs, the topmost GaAs quarter-wave anti-phase layer is etched

6 Improved and Alternative Atomic Clock VCSELs

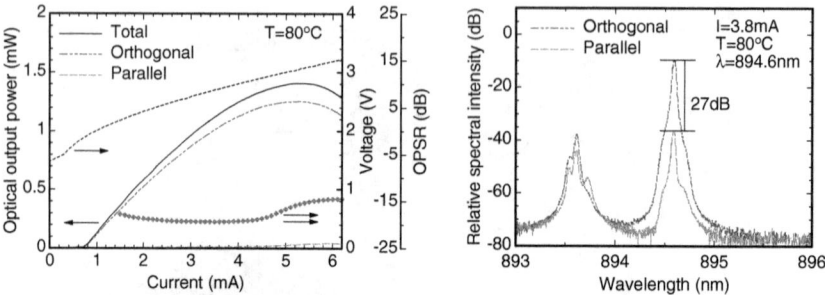

Figure 6.10: Polarization-resolved operation characteristics of a grating relief VCSEL with $D_\mathrm{a} = 4.5\,\mathrm{\mu m}$, measured at $T = 80°\mathrm{C}$ (left) and its polarization-resolved spectra at $I = 3.8\,\mathrm{mA}$ (right). The grating relief has a diameter of $3.3\,\mathrm{\mu m}$, a grating period of $0.6\,\mathrm{\mu m}$ and an etch depth of $70\,\mathrm{nm}$. The laser has $x = 4\%$ and $N_\mathrm{p} = 25$.

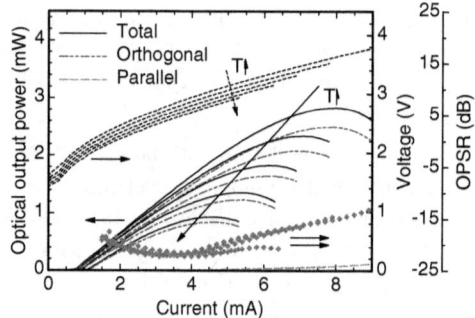

Figure 6.11: Polarization-resolved operation characteristics of a grating relief VCSEL with $D_\mathrm{a} = 5\,\mathrm{\mu m}$, measured at T varied between 20 and $100°\mathrm{C}$ in steps of $20°\mathrm{C}$. The grating relief has a diameter of $4\,\mathrm{\mu m}$, a grating period of $0.6\,\mathrm{\mu m}$ and an etch depth of $70\,\mathrm{nm}$. The laser has $x = 4\%$ and $N_\mathrm{p} = 25$.

over the whole outcoupling facet. This means that in-phase reflection is achieved for all transverse modes. The reference VCSELs can be thus considered as standard VCSELs. Figure 6.12 displays the PR-LI characteristics at $T = 80°\mathrm{C}$ of a reference device with an oxide aperture of about $5\,\mathrm{\mu m}$ and its optical spectra at different driving currents. The laser has a threshold current of $I_\mathrm{th} = 0.7\,\mathrm{mA}$ and a maximum output power of $4.2\,\mathrm{mW}$. At $1.5\,\mathrm{mA}$ drive current it shows single-mode operation with an SMSR of $20\,\mathrm{dB}$. However, the spectrum gets highly multi-mode for higher currents. Having no surface grating, the reference VCSEL shows a weak polarization control with an average OPSR of $-4.1\,\mathrm{dB}$.

6.2 Alternative Surface Grating Approaches

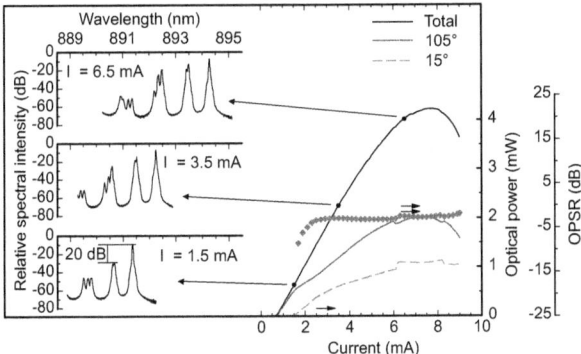

Figure 6.12: Polarization-resolved operation characteristics of a reference VCSEL with $D_a = 5\,\mu\text{m}$, measured at $T = 80°\text{C}$. The emission spectra in the insets show higher-order lasing modes. The polarization directions of the two orthogonal, linearly polarized fundamental modes are rotated by 15° towards the $[0\bar{1}1]$ axis. The laser has $x = 4\%$ and $N_\text{p} = 25$.

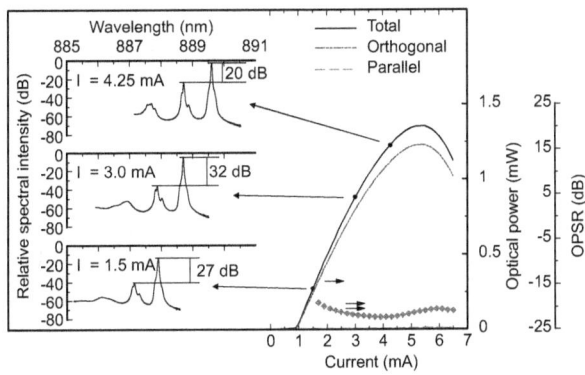

Figure 6.13: Polarization-resolved operation characteristics of the grating relief VCSEL from Fig. 6.11 at 80°C substrate temperature. The emission spectra in the insets show SMSRs of at least 20 dB.

The OPSR is calculated for data points in steps of 0.1 mA and then averaged over the current range yielding 10 % to 100 % of the maximum output power. Due to built-in strain forces, the two orthogonal, linearly polarized fundamental modes are not aligned parallel and orthogonal to the usually preferred [011] crystal axis. Instead, they are rotated by 15° towards the $[0\bar{1}1]$ axis because of the elasto-optic effect, which was briefly explained in

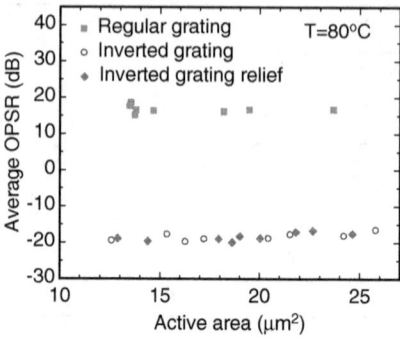

Figure 6.14: Average OPSR versus active area like Fig. 6.6 with data of several inverted grating relief VCSELs added. $T = 80°C$, $x = 4\%$ and $N_\text{p} = 25$ for all devices.

Sect. 3.6. Figure 6.13 depicts the same measurements for a nearby laser (same as Fig. 6.11) on the same sample, which is nominally identical except for a surface grating relief with a diameter of 4 µm. The grating relief device shows an increased threshold current of 0.9 mA due to the effectively decreased mirror reflectivity. The optical spectra confirm SMSRs exceeding 20 dB up to 4.25 mA at which the laser delivers a maximum single-mode output power of 1.2 mW. This current is just 1.25 mA below the thermal roll-over point. Owing to the grating, the VCSEL is polarization-stable well above thermal roll-over with an average OPSR of −21 dB.

To compare the polarization control induced by the inverted grating reliefs with the full-area surface gratings, e.g., the regular grating and inverted grating VCSELs shown in Fig. 6.6, average OPSRs of several grating relief VCSELs have been experimentally obtained and added (as green symbols) to Fig. 6.14. The grating reliefs have diameters ranging between 3 and 4 µm, but fixed etch depths of 70 nm and grating periods of 0.6 µm. They show negative OPSRs with absolute values comparable to those of inverted grating VCSELs (blue circles). The dominant polarization modes are always orthogonal to the grating lines.

The effect of the grating relief on the beam properties is investigated by measuring the emission far-fields. Figure 6.15 shows the far-fields orthogonal and parallel to the grating lines of a grating relief VCSEL with 5 µm active diameter, measured at $I = 2$ mA and $T = 20°C$. The grating relief has a diameter of 3 µm, a grating period of 0.6 µm, and an etch depth of 70 nm. Showing no side-lobes in emission far-fields proves the absence of diffraction losses in air. This is a favorable functional property of the sub-emission-wavelength grating periods. Almost identical far-field patterns along the orthogonal and

6.2 Alternative Surface Grating Approaches

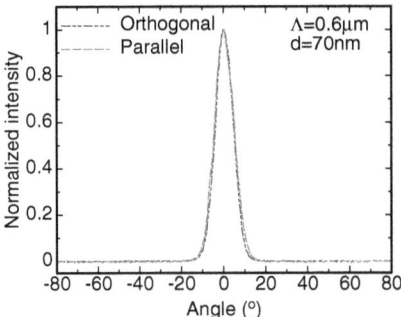

Figure 6.15: Emission far-fields of a grating relief VCSEL with $D_a = 5\,\mu\mathrm{m}$, measured at $T = 20\,^\circ\mathrm{C}$. The almost overlapping dash-dotted and dashed lines indicate the normalized optical intensity orthogonal and parallel to the grating lines, respectively. The laser has $x = 4\%$ and $N_p = 25$.

parallel directions indicate a circular beam profile. The Gaussian-like shape reflects the single-mode emission of the VCSEL at the given current, which proves the functionality of grating reliefs in enhancing the fundamental-mode emission for VCSELs with relatively large active diameters (e.g., $D_a = 5\,\mu\mathrm{m}$). The values of θ_0, θ_{FWHM} and w_0 obtained by a Gaussian curve fit (not shown here but similar to what was done earlier in Figs. 5.7 and 6.8) are 8.8°, 10.4° and 1.85 µm, respectively. The diameter of the beam waist $2w_0 = 3.7\,\mu\mathrm{m}$ is slightly smaller than the active diameter $D_a = 5\,\mu\mathrm{m}$. This may indicate that the grating relief contributes to the waveguiding in the VCSEL.

The above results show that the grating relief and the full-area grating techniques have comparable polarization control and far-field properties. However, the grating relief technique results in polarization-stable single-mode VCSELs with larger D_a reaching up to 5 µm, compared to standard small-aperture single-mode devices with $D_a = 3$ to 4 µm. Such larger active diameters can provide greater potential for increased device lifetime. This might be desired for some MAC applications. Preliminary reliability tests of such devices will be shown in Sect. 6.4. As a drawback, the grating relief technique leads to larger threshold currents compared to the full-area surface grating techniques. However, such a problem could be solved by increasing the number of top mirror pairs, as discussed in Sect. 6.1.

6.3 Reduction of Processing Complexity

The dynamic characteristics presented in Sect. 5.2 were obtained with VCSEL chips fabricated at the beginning of the research performed for this dissertation. These chips have wet-chemically-etched p-mesas with diameters D_m at the oxidation layer of about 43 µm and polyimide thicknesses of about 7.5 µm. In the current section, dynamic characteristics of more recent VCSEL chips with smaller wet-chemically-etched or dry-etched p-mesas as well as thinner polyimide layers will be presented. As an example, a VCSEL has an active diameter of 5.3 µm, a wet-chemically-etched p-mesa with a diameter of 35 µm and a polyimide thickness of 5.5 µm[3]. Figure 6.16 (top) depicts the measured S_{11} spectra of the VCSEL chip on a Smith chart at different bias points above threshold at $T = 20°C$. The real and imaginary parts of the S_{11} data are plotted in Fig. 6.16 (bottom).

Applying the electrical equivalent-circuit model in Fig. 5.11, the values of the parasitic elements shown in Table 6.1 could be extracted from curve fits which are plotted (as solid curves) in Fig. 6.16 (top) and (bottom) using (5.10) and (5.6). As expected, L, R_m, and C_{pad} are independent of I. The observed increase of C_a with increasing I is a consequence of the increase of C_{diff} from (5.5). The decrease of R_a is consistent with the decrease of the differential resistance in the IV curve at higher bias. The current-dependent behavior of R_a and C_a results in a small increase of f_p with I. In comparison with the parasitic elements extracted for VCSEL chips presented earlier in Chap. 5, like the one of Table 5.1, one observes a decrease of C_a by approximately a factor of 1.6. This is due to differences in mesa size and in active diameter, resulting in a decrease of the area of the oxide layer by almost the same factor[4], which causes reductions of C_{ox} and C_{dep} in (5.5). The somewhat larger active diameter of the VCSEL from Table 6.1 results in a smaller R_a. Such a reduction of the RC product improves the electrical parasitic 3 dB corner frequency f_p by approximately a factor of 1.9 to about 10 GHz.

In order to reduce the fabrication complexity, VCSELs with reduced polyimide thickness according to Fig. 4.22 were also fabricated. Mesas with smaller diameters are employed in these VCSELs, resulting in a smaller area of the oxide layer and hence reduced C_{ox} and

---

[3] As explained in Sect. 4.6.2, there is no reduction of the process complexity for the VCSELs with 5.5 µm polyimide thickness compared to the VCSELs with 7.5 µm polyimide thickness. The former is wire-bondable and requires an additional final annealing step at 330°C. Moreover, the preceding polyimide passivation layer is hard-baked at slightly higher temperature (e.g., 350°C instead of 300°C).

[4] For instance, at $I = 2\,\text{mA}$, C_a of the VCSELs from Tables 5.1 and 6.1 decreases from 392 to 238.5 fF, which is a reduction by almost a factor of 1.65. Comparing the oxide areas A_{ox1} and A_{ox2} of the VCSELs from Tables 5.1 and 6.1, respectively, one gets $A_{ox1} = \pi(D_m^2 - D_a^2)/4 = \pi((43\,\text{µm})^2 - (3.6\,\text{µm})^2)/4 = 1442\,\text{µm}^2$ and $A_{ox2} = \pi((35\,\text{µm})^2 - (5.3\,\text{µm})^2)/4 = 940\,\text{µm}^2$, then the ratio $A_{ox1}/A_{ox2} = 1.55$.

6.3 Reduction of Processing Complexity

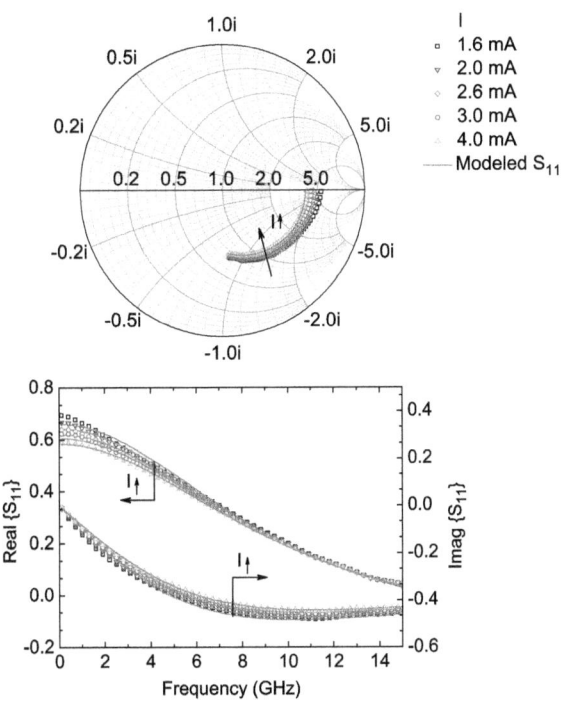

Figure 6.16: Measured S_{11} spectra of a VCSEL chip with 5.3 µm active diameter at 1.6, 2, 2.6, 3 and 4 mA bias currents and $T = 20°C$ on a Smith chart over a frequency range from 0.1 to 15 GHz in 100 MHz steps (top) and their real and imaginary parts (bottom). The solid lines are fits to the measurement data and represent the modeled S_{11} using the equivalent-circuit depicted in Fig. 5.11. The laser has $x = 4\%$ and $N_\mathrm{p} = 25$.

C_dep. C_a from (5.5) is consequently also reduced. Moreover, to further decrease the oxide area, dry-etched mesas with vertical sidewalls are employed rather than wet-chemically-etched mesas with inclined walls (see Fig. 4.20). A thick polyimide VCSEL (Fig. 4.18 (right) and Tables 6.1 and 5.1) requires i) two mesa etching steps with different diameters which must reach the highly n-doped contact layer and ii) the deposition of three polyimide layers by means of three lithographic steps. On the other hand, a thin polyimide VCSEL (Fig. 4.22 and Table 6.2) requires just one mesa etching step down to, at least, the first layer pair of the n-type DBR and a single polyimide passivation layer. However, variations of the etch depth in the n-type DBR (as explained in Sect. 4.6.2) cause variations of the

6 Improved and Alternative Atomic Clock VCSELs

Table 6.1: Extracted values of the equivalent-circuit elements and their 3 dB corner frequencies at different I and $T = 20°C$ for the VCSEL chip from Fig. 6.16. The active diameter is 5.3 µm, the p-mesa diameter is 35 µm and the polyimide thickness is 5.5 µm.

I (mA)	L (pH)	R_m (Ω)	R_a (Ω)	C_{pad} (fF)	C_a (fF)	f_p (GHz)
1.6	95	52	192	82	233.5	9.6
2.0	95	52	177	82	238.5	9.65
2.6	95	52	161	82	244.0	9.8
3.0	95	52	151	82	248.0	9.85
4.0	95	52	138	82	252.5	10.1

Table 6.2: Equivalent-circuit elements like in Table 6.1 for a thin polyimide VCSEL chip at $T = 20°C$. The active diameter is 5.2 µm, the p-mesa diameter is 26 µm and the polyimide thickness is 1.5 µm.

I (mA)	L (pH)	R_m (Ω)	R_a (Ω)	C_{pad} (fF)	C_a (fF)	f_p (GHz)
1.6	95	54	207	138	215	9.1
2.0	95	54	188	138	223	9.1
2.6	95	54	160	138	229	9.4
3.0	95	54	155	138	233	9.4
4.0	95	54	140	140	245	9.4

doping level at the surface, given that modulation doping is applied in the DBR. Therefore, fluctuations of the electrical resistance of the Bragg mirror R_m might appear for different devices depending on the position from the wafer center. Characterizations performed in this dissertation did not show significant fluctuations of R_m.

Similarly to what was done for the VCSEL from Fig. 6.16, electrical parasitic components of the equivalent-circuit model of a thin polyimide VCSEL with 5.2 µm active diameter having a 1.5 µm thick polyimide layer and a 26 µm diameter dry-etched p-mesa have been determined and listed in Table 6.2 for different bias currents. The extracted elements have a similar current dependence as described for Table 6.1. However, a significant increase of C_{pad} by approximately a factor of 1.7 is observed for this VCSEL[5]. Differences in mesa size and in active diameter account for a decrease of the area of the oxide layer by

---

[5] Unexpectedly, the VCSELs with 5.5 µm and 1.5 µm polyimide thickness from Tables 6.1 and 6.2, respectively, show smaller C_{pad} compared to the VCSELs with thicker polyimide layers (e.g., 7.5 µm) from Tables 5.1 and 5.2. No simple explanation of such behavior has been found. It might indicate a general limitation of the proposed equivalent-circuit model of Fig. 5.11.

6.3 Reduction of Processing Complexity

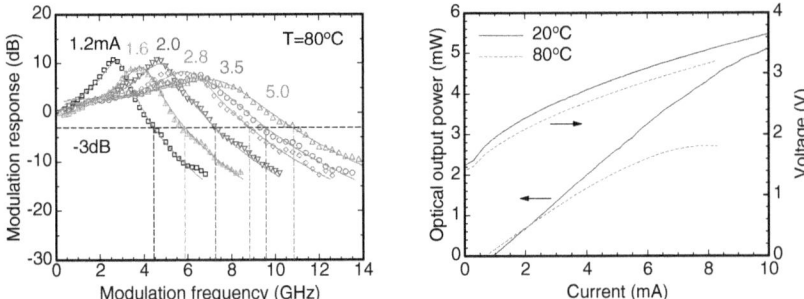

Figure 6.17: Small-signal modulation response curves of the VCSEL from Table 6.2 along with curve fits as solid lines according to (3.33) at different bias currents and $T = 80°C$ (left), and LIV characteristics of the same device at $T = 20$ and $80°C$ (right). The laser has $x = 4\%$ and $N_p = 25$.

approximately a factor of 1.8. This reduces C_{ox} and C_{dep} in (5.5) and, to a smaller degree, also C_a. The about equal active diameters of the VCSELs from Table 6.1 and Table 6.2 result in approximately equal R_a. The purpose of using a thick polyimide layer (Tables 6.1 and 5.1 with 5.5 μm and 7.5 μm polyimide thickness, respectively) was to reduce C_{pad} and thus to improve the parasitic bandwidth. Nevertheless, the parasitic corner frequency f_p of the thin polyimide VCSEL (Table 6.2 with 1.5 μm polyimide thickness) is just smaller by 0.5 to 0.7 GHz, where the f_p of both VCSELs are above 9 GHz and thus 4 to 5 GHz higher compared to the larger p-mesa VCSELs introduced in Sect. 5.2. Therefore, instead of increasing the polyimide thickness, almost similar parasitic behavior is achieved with a single polyimide step and a simultaneous reduction of the mesa size. Clearly, the processing of thin polyimide VCSELs is much less complex and more desired for large production volumes.

To fully characterize the dynamic behavior of thin polyimide VCSELs, their small-signal modulation responses and intrinsic modulation bandwidths are introduced in what follows. Small-signal modulation response curves of the VCSEL from Table 6.2 are presented in Fig. 6.17 (left) for $T = 80°C$ at different bias currents. Figure 6.17 (right) displays the LIV characteristics of the VCSEL at $T = 20$ and $80°C$. The threshold current is 0.9 mA at $20°C$ and only 0.7 mA at $80°C$, which is the result of the choice of $x = 4\%$ In content in the QWs (see Sect. 4.2.6). For polarization control, the VCSEL has an inverted grating relief. It shows thus relatively higher threshold currents compared to the full-area grating VCSELs (e.g., devices from Fig. 6.5). As usual it is observed that at higher bias currents, the resonance frequency f_r and the damping coefficient γ increase, thus the resonance peak becomes more damped and shifts to higher frequencies. The maximum 3 dB

117

6 Improved and Alternative Atomic Clock VCSELs

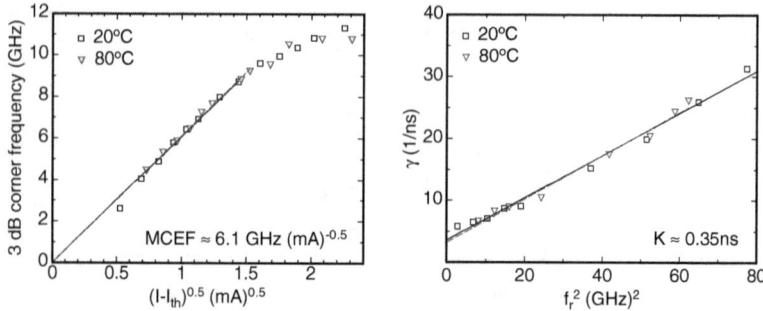

Figure 6.18: 3 dB corner frequency as a function of $(I - I_{\text{th}})^{0.5}$ for the VCSEL from Fig. 6.17 along with linear fits at $T = 20°\text{C}$ (solid) and $80°\text{C}$ (dashed) with the MCEFs determined from their slopes (left). Damping coefficient γ versus resonance frequency f_{r} squared of the same VCSEL along with linear fits at $T = 20°\text{C}$ (solid) and $80°\text{C}$ (dashed) with the K-factors taken from their slopes (right).

bandwidth of 10.8 GHz is obtained at $I = 5.0$ mA. A bandwidth of 5.85 GHz, already exceeding the target value of 5 GHz, is obtained at $I = 1.6$ mA, i.e., only 0.9 mA above threshold. In order to determine the MCEF, 3 dB corner frequencies of the VCSEL are extracted from the small-signal modulation responses and plotted against $\sqrt{I - I_{\text{th}}}$ in Fig. 6.18 (left). Equivalent measurements of modulation responses of the same VCSEL were also performed at 20°C and the $f_{\text{3 dB}}$ are included in Fig. 6.18 (left). The MCEFs are determined from linear fits. A value of approximately $6.1\,\text{GHz}/\sqrt{\text{mA}}$ is obtained for both temperatures. Using the modulation transfer function (3.33) and the extracted values of f_{p}, precise fits of the measured modulation response curves can be performed and f_{r} and γ can be extracted. Figure 6.17 (left) shows the fit curves (lines) superimposed on the experimental data. The measurements and the extraction of parasitics were also performed for $T = 20°\text{C}$. From the fit parameters, the K-factor is then determined as the slope of the linear fit of γ versus f_{r}^2, as shown in Fig. 6.18 (right). It has a value of about 0.35 ns for both temperatures. The damping-limited maximum 3 dB corner frequency $f_{\text{max,d}}$ according to (3.41) is 25.4 GHz.

6.4 Reliability Tests

For preliminary reliability testing, a sample containing several inverted grating relief VCSELs was introduced in a setup in which six individual lasers with 5.5 (2 devices), 5.0, 4.4, 3.8 and 3.7 µm active diameter are operated at a constant current of 3 mA. The

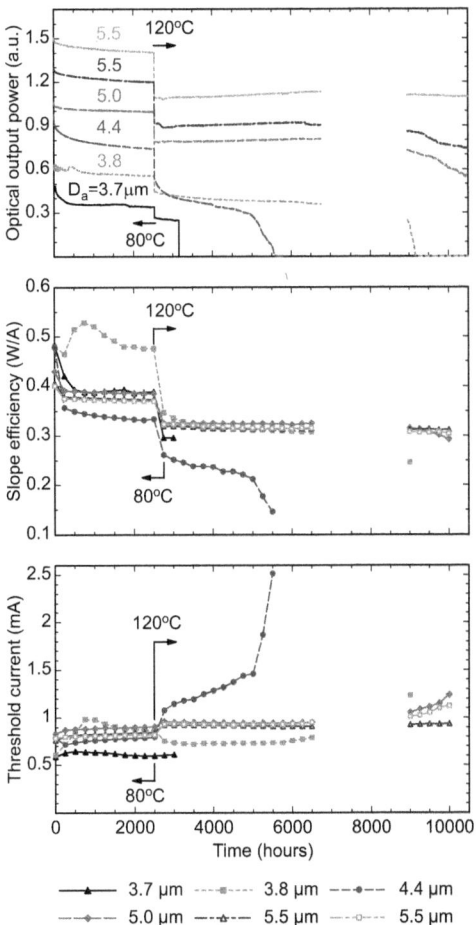

Figure 6.19: Output power (top), slope efficiency (middle) and threshold current (bottom) evolution in a long-term test of grating relief VCSELs with different active diameter D_a, all driven at 3 mA bias current. After 2500 hours the ambient temperature was increased from 80 to 120°C.

optical output power P of each device is measured separately (while the other devices are switched off) using a 1 mm² area silicon photodetector. The power was recorded using a personal computer (PC) every half hour for about 10000 hours. Unfortunately, due to a technical problem of the PC, data points between the hour 6700 and 9000 have not been

6 Improved and Alternative Atomic Clock VCSELs

Table 6.3: Current density, thermal resistance, internal temperature and aging acceleration factor for the VCSELs from Fig. 6.19 at $T = 80°$C. J, T_{int} and x_{aging} are obtained for $I = 3$ mA.

D_{a} (µm)	J (kA/cm^2)	R_{th} (K/mW)	T_{int} (°C)	x_{aging}
5.5 (2 devices)	12.6	3.3	101	1
5.0	15.3	3.6	103	1.7
4.4	19.7	4.1	107	3.4
3.8	26.5	4.75	111	7.8
3.7	27.9	4.9	116	11.3

recorded. Figure 6.19 (top) shows P versus time for all devices. Despite the interruptions of the curves, their behavior can still be anticipated by simple extrapolations. Obviously such a small number of devices is not sufficient to obtain reliable lifetime estimations. Still it can be stated that the lasers showed only minor aging after 2500 hours at $T = 80°$C. Owing to the different active diameters, the current density J of the six VCSELs is not constant. J increases for decreased D_{a}, as can be seen in Table 6.3.

The laser lifetime decreases with increasing J and increased internal laser temperature T_{int}. It is usually characterized in terms of the mean time to failure (MTTF) which is given by the empirical relationship [205]

$$\text{MTTF} \propto \frac{\exp\{E_{\text{a}}/(k_{\text{B}} T_{\text{int}})\}}{J^2}, \tag{6.1}$$

where k_{B} is Boltzmann's constant, T_{int} is the average temperature (in Kelvin) of the inner cavity of the VCSEL and E_{a} is the failure activation energy taking values of about 0.7 eV for commercial single-mode VCSELs [191]. MTTF is the predicted average time for a laser being operated at a given J and T before becoming defective. Often an output power decay by 2 dB is rated as a failure. The thermal resistance R_{th} of the VCSELs increases for decreased D_{a}, causing increased T_{int} and hence more stressful and accelerated aging. The values of R_{th}, T_{int} and J for the six VCSELs are listed in Table 6.3 for $T = 80°$C. T_{int} was calculated using (3.26), where P_{diss} has been obtained using (3.27) at $I = 3$ mA after 2500 hours of operation.

An aging acceleration factor x_{aging} relative to a reference VCSEL (e.g., the VCSEL with the largest $D_{\text{a}} = 5.5$ µm) can be derived from (6.1) as

$$x_{\text{aging}} = \left(\frac{J}{J_0}\right)^2 \cdot \exp\left\{\frac{E_{\text{a}}}{k_{\text{B}}}\left(\frac{T_{\text{int}} - T_{\text{int},0}}{T_{\text{int}} T_{\text{int},0}}\right)\right\} \tag{6.2}$$

6.4 Reliability Tests

with J_0 and $T_{\text{int},0}$ being the current density and the average internal temperature (in Kelvin) of the reference VCSEL, respectively. x_{aging} relative to the VCSEL with $D_\text{a} = 5.5\,\mu\text{m}$ is shown in Table 6.3, where E_a is assumed to be 0.7 eV. A maximum x_{aging} of 11.3 is obtained for the VCSEL with the smallest active aperture $D_\text{a} = 3.7\,\mu\text{m}$. In other words, the MTTF of this VCSEL is expected to be shorter than the reference VCSEL by a factor of 11.3 when both are operated at $I = 3\,\text{mA}$ and $T = 80°\text{C}$. x_{aging} is mainly governed by the different J and to a smaller degree by a varying T_{int}.

For accelerated aging of the entire set of VCSELs, T was increased to 120°C, which can be seen as an abrupt decrease of P. This is mainly due to thermally-induced leakage currents and reduced material gain. The devices with $D_\text{a} = 3.7$, 3.8 and 4.4 μm which showed much initial power reduction or fluctuations at 80°C became unfunctional after 620, 6660 and 3000 hours of operation at 120°C, respectively. The other lasers with $D_\text{a} \geq 5\,\mu\text{m}$ kept to be functional with slower degradation. For more insight into the operation characteristics of the tested lasers, slope efficiency SE and threshold current I_{th} versus time are depicted in the middle and bottom graphs of Fig. 6.19, respectively. The step-increase from $T = 80$ to 120°C appears as a decrease in SE by 13.5% (2 devices), 13%, 21%, 27% and 23% for the VCSELs with $D_\text{a} = 5.5$ (2 devices), 5, 4.4, 3.8 and 3.7 μm, respectively. For all lasers, I_{th} slightly increases ($< 13.5\%$) after the step-increase in T except for the VCSELs with $D_\text{a} = 4.4$ and 3.8 μm. I_{th} of the first increased by 35% and became higher than 1 mA, then kept increasing to approximately 2.5 mA before becoming unfunctional at the hour 5500, while for the latter it decreased by 13%, then kept increasing to approximately 1.2 mA before becoming unfunctional at the hour 9160. The voltages V remained almost unchanged except for the laser with $D_\text{a} = 3.7\,\mu\text{m}$ whose V and differential series resistance started to increase at the hour 1500 from 2.59 V and 343 Ω to 2.82 V and 412 Ω, respectively, at the hour 2500. From the unstable behavior of P, I_{th}, SE, and V over time we can expect that lasers with $D_\text{a} \geq 5\,\mu\text{m}$ provide greater potential for increased lifetime in comparison with standard small-aperture single-mode devices with $D_\text{a} = 3$ to 4 μm.

Chapter 7

Experimental Cesium-Based Atomic Clock Demonstrator

This chapter reports about investigations on special characteristics of atomic clock VCSELs performed at UniNE-LTF and FEMTO-ST[1]. Such characteristics include emission spectral linewidth, relative intensity and frequency noises in the low-frequency range as well as modulation sideband characteristics. They demonstrate the high-level performance of the VCSELs and prove their validity and suitability as laser sources for highly-stable cesium-based MACs. Due to limitations in time, a small yield of atomic clock sub-systems (electronic circuitries, Cs vapor cells, ...) and integration difficulties, only a few MAC demonstrators have been assembled within the duration of the MAC-TFC project. Unfortunately, such demonstrators suffered from some defects and showed unexplained low-contrast CPT resonance signals. As a backup, an experimental larger-scale atomic clock demonstrator has been built at FEMTO-ST using miniaturized components which were fabricated in the first place for the MAC demonstrator. The experimental demonstrator has been successfully operated, showing CPT resonance signals with narrow linewidth, as will be illustrated at the end of this chapter.

7.1 VCSEL Description and Packaging

Two different VCSEL designs have been utilized for the experimental results introduced in this chapter. Both have a layer structure similar to the one introduced in Sect. 4.5 and have a top annular p-contact but either a substrate side or a top n-contact. The latter has the flip-chip-bondable design described in Sect. 4.6.1. In addition, it applies the inverted grating relief technique by which favorable single-mode and polarization-stable

---

[1] UniNE-LTF and FEMTO-ST are MAC-TFC project partners. See Sect. 2.3 and App. B to know more about the MAC-TFC consortium.

7 Experimental Cesium-Based Atomic Clock Demonstrator

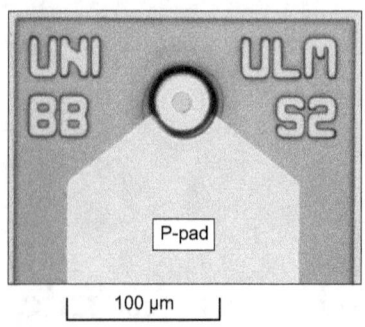

Figure 7.1: Photograph of a fully processed VCSEL with a standard n-type substrate-side contact, suitable for wire bonding of the p-contact.

laser emission is achieved, as explained and experimentally illustrated in Sect. 4.4.2 and Sect. 6.2.2, respectively. The surface grating is etched in an extra topmost GaAs quarter-wave anti-phase layer. The grating relief and its surface profile measured with an AFM was displayed by Fig. 6.9 (b) and (c).

VCSELs with substrate-side n-contact were processed within an earlier research activity before the start of the present dissertation work. A detailed fabrication and processing description of such devices can be found in Ref. [138]. One can refer to such a design as a *standard VCSEL*. Figure 7.1 shows such a VCSEL chip having $250 \times 250\,\mu m^2$ size.

7.1.1 Standard VCSELs

Several standard VCSELs have been mounted in TO-46 cans for convenient testing. Having substrate-side n-contacts, the laser chips were fixed on silicon submounts by conductive glue. The anode of the TO-46 can is wire-bonded to the top p-pad while the cathode is wire-bonded to the Au-metallized top of the silicon submount[2]. An external lens is employed for laser beam collimation. A thermoelectric cooler (TEC) and a thermistor are fixed in proximity of the VCSEL package for temperature control. Having no surface gratings, the polarization of the light emitted from standard VCSELs is a priori undefined, as explained in Sect. 3.6. Devices with stable polarization in the operation range of interest have been selected. The PR-LIV characteristics of such a laser are shown in Fig. 7.2 (left). Single-mode emission polarized parallel to the [011] GaAs reference crystal axis is maintained up to a current of 4.4 mA. Over this range, the OPSR is greater than 20 dB, where the OPSR is calculated according to (5.1), but here P_{par} and P_{orth} are the optical powers measured behind a Glan–Thompson polarizer whose transmission direction is oriented parallel and orthogonal to the [011] axis, respectively.

---

[2]These mounting steps, except wire-bonding, were performed at Philips U-L-M Photonics in Ulm.

7.1 VCSEL Description and Packaging

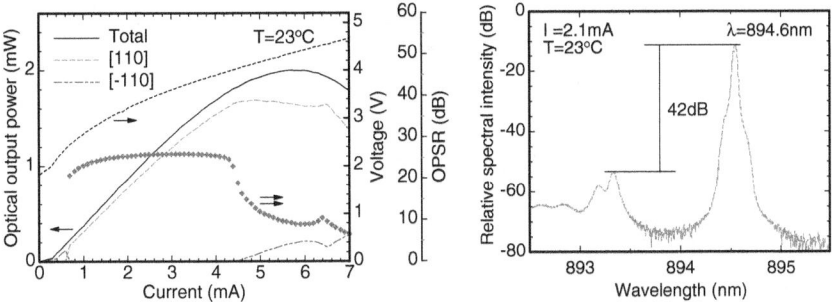

Figure 7.2: Polarization-resolved operation characteristics of a standard VCSEL with $D_a = 3\,\mu\text{m}$ at $T = 23°C$ (left) and its spectrum at $I = 2.1\,\text{mA}$ (right).

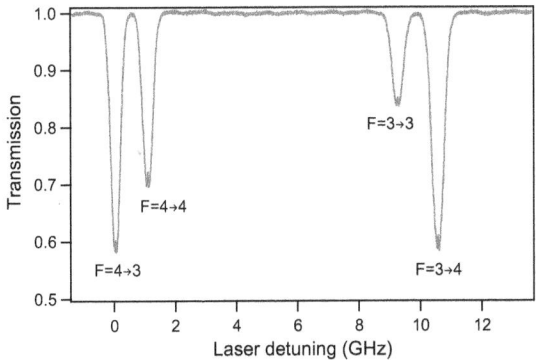

Figure 7.3: Cs D_1 absorption lines observed when transmitting the light of the VCSEL from Fig. 7.2 through a Cs vapor cell (figure courtesy of UniNE-LTF) [206].

Figure 7.2 (right) depicts the optical spectrum at $23°C$. The Cs D_1 line wavelength of $894.6\,\text{nm}$ is reached at a current of $2.1\,\text{mA}$. The SMSR is about $42\,\text{dB}$. This laser was utilized to resolve all four hyperfine components of the Cs D_1 line, which are clearly separated in Fig. 7.3. The fine and hyperfine structures of the energy level system of Cs are explained and illustrated in App. A.1. Cs atomic transitions are indicated using the quantum number F associated with the total atomic angular momentum $\mathbf{F}$, as explained in App. A.1. The laser frequency is detuned over a $14\,\text{GHz}$ span at a rate of $\approx 300\,\text{MHz}/\mu\text{A}$ by sweeping the laser current. The VCSEL was operated at the bias current and temperature from Fig. 7.2 (right). The Cs absorption maxima appearing at detuning frequencies of nearly 0, 1.2, 9.2 and $10.4\,\text{GHz}$ correspond to the $|F=4\rangle \to |F'=3\rangle$, $|F=4\rangle \to |F'=4\rangle$,

125

 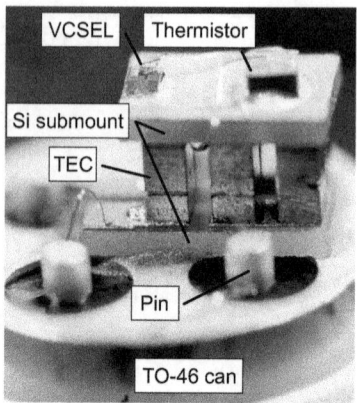

Figure 7.4: VCSEL mounted along with a thermistor and a TEC in a TO-46 can (left). A close-up of the mount assembly taken at a different angle (right). The can has 5 pins: one for the thermistor, one for the p-pad of the VCSEL, two for cooling and heating of the TEC and one as a common ground for all elements. The ground pin is below the lower Si submount and cannot be seen.

$|F = 3\rangle \rightarrow |F' = 3\rangle$ and $|F = 3\rangle \rightarrow |F' = 4\rangle$ atomic transitions, respectively, as indicated in Fig. 7.3. F is designated for the hyperfine structure of the $6^2S_{1/2}$ ground level and F' is assigned to the hyperfine structure of the $6^2P_{1/2}$ excited level, as depicted in Fig. A.1. Not well-resolved saturated absorption features can be seen in the four lines. However, better resolved saturation features were obtained, as will be shown in Fig. 7.6.

7.1.2 Inverted Grating Relief VCSELs

Several flip-chip-bondable inverted grating relief VCSELs according to the design shown in Fig. 6.9 with the simple cross-section depicted in Fig. 4.22 have been mounted in TO-46 cans. Having top n-contacts, such VCSELs were fixed on silicon submounts using glue. The anode and cathode of the TO-46 can were wire-bonded to the top p- and n-pads of the VCSEL. For temperature control, a TEC and a thermistor are placed inside the can (here, the mounting was entirely done at U-L-M, see footnote 2 on p. 124). Figure 7.4 displays such a mount assembly.

PR-LIV characteristics of such a VCSEL at $T = 30°C$ are shown in Fig. 7.5 (left). Five times P_{par} is plotted here for better clarity. The device remains polarization-stable up to thermal roll-over with a maximum magnitude of the OPSR of 22.7 dB. The polarization of inverted grating VCSELs with optimum design is always orthogonal to the grating lines,

7.2 Laser Noise and Dynamics

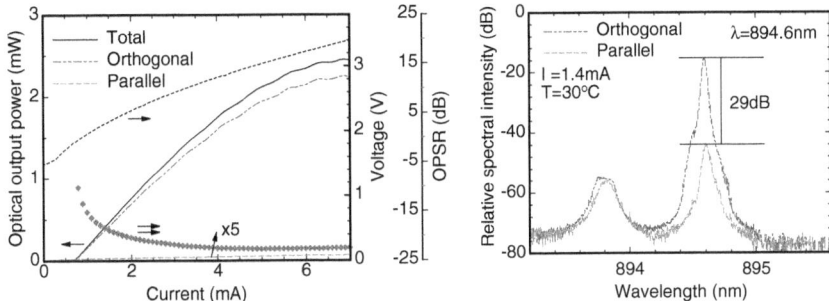

Figure 7.5: Polarization-resolved operation characteristics of an inverted grating relief VCSEL with $D_\mathrm{a} = 3.6\,\mu\mathrm{m}$ at $T = 30°\mathrm{C}$ (left) and its polarization-resolved spectra at $I = 1.4\,\mathrm{mA}$ (right). The grating relief has a diameter of $3\,\mu\mathrm{m}$, a grating period of $0.6\,\mu\mathrm{m}$ and an etch depth of $70\,\mathrm{nm}$.

resulting in an opposite sign of the OPSR in Fig. 7.5 (left) compared to Fig. 7.2 (left). Figure 7.5 (right) depicts polarization-resolved spectra at 30°C. The target wavelength is reached at a current of $1.4\,\mathrm{mA}$ with an SMSR of about $40\,\mathrm{dB}$ and a peak-to-peak difference between the dominant and the suppressed polarization modes of almost $29\,\mathrm{dB}$.

7.2 Laser Noise and Dynamics

7.2.1 Emission Linewidth

Standard VCSELs show threshold currents of around $0.3\,\mathrm{mA}$ (see Fig. 7.2 (left)) with weak dependence on temperature, namely less than $2\,\mu\mathrm{A/K}$ in a range of 15°C to 25°C. As seen from Fig. 7.2, the Cs D_1 line wavelength is reached at a bias current of $2.1\,\mathrm{mA}$ and a temperature of 23°C, with an output power of $910\,\mu\mathrm{W}$ and more than $40\,\mathrm{dB}$ SMSR. Figure 7.6 shows an emission linewidth spectrum of the VCSEL from Fig. 7.2. The spectrum is measured using a scanning Fabry–Pérot interferometer with $5\,\mathrm{MHz}$ intrinsic linewidth and $1\,\mathrm{GHz}$ free spectral range. Typical measured linewidths are around $\Delta f_\mathrm{L} = 20\text{--}25\,\mathrm{MHz}$, by which narrow saturated absorption lines can be resolved, as depicted in the inset of Fig. 7.6. Similar saturated absorption features were resolved for the Cs D_2 line at $852\,\mathrm{nm}$ wavelength [207]. Such saturated features show that the emission linewidth of VCSELs is narrower than the linewidths of the Cs absorption lines. Therefore, VCSELs are suitable laser sources for CPT-based atomic clocks in spite of their relatively large linewidths compared to other semiconductor lasers such as DFB or external-cavity

7 Experimental Cesium-Based Atomic Clock Demonstrator

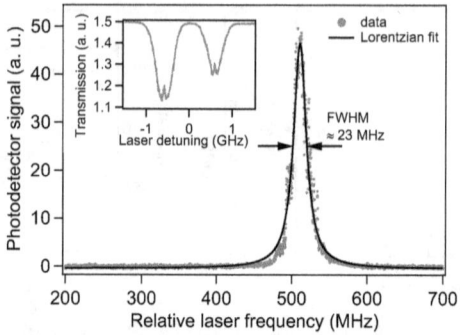

Figure 7.6: Emission linewidth spectrum of the VCSEL from Fig. 7.2 operating at the Cs D_1 line. The Fabry–Pérot linewidth is 5 MHz and the total sweep time for this graph is 30 ms. The inset shows narrow saturated absorption features for the hyperfine transitions $|F = 4\rangle \to |F' = 3\rangle$ and $|F = 4\rangle \to |F' = 4\rangle$ of the Cs D_1 line, which were recorded for an evacuated Cs vapor cell (figure courtesy of UniNE-LTF) [206].

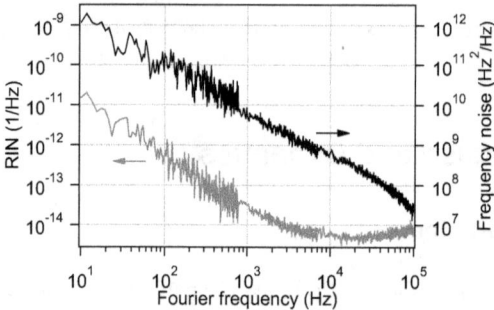

Figure 7.7: RIN and frequency noise of the VCSEL from Fig. 7.2 operating at the Cs D_1 line in a free-running mode (figure courtesy of UniNE-LTF) [206].

laser diodes. An independent determination of the VCSEL linewidth can be obtained from the spectral power density of its frequency noise $\tilde{S}_{FN}(f)$ shown in Fig. 7.7. As mentioned in Sect. 3.5.5, the VCSEL linewidth can be estimated by (3.62) resulting in $\Delta f_L \approx 20$ MHz. This value is consistent with linewidths measured by the Fabry–Pérot interferometer.

7.2 Laser Noise and Dynamics

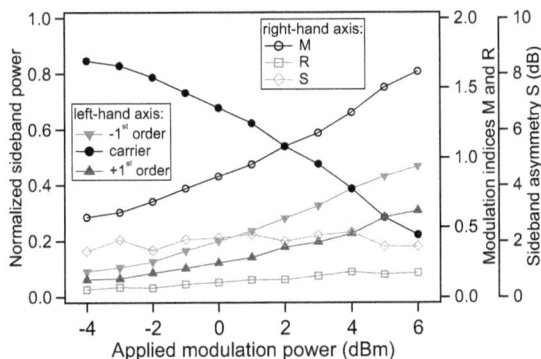

Figure 7.8: Normalized power of the carrier and the first-order sidebands (filled symbols, left-hand axis) and corresponding modulation indices M and R and sideband asymmetry factor S (open symbols, right-hand axis) versus the RF modulation power with $f = 4.596\,\text{GHz}$ applied to the VCSEL from Fig. 7.2 (figure courtesy of UniNE-LTF) [206].

7.2.2 Relative Intensity and Frequency Noise

Figure 7.7 depicts RIN and frequency noise of a standard VCSEL emitting at the Cs D_1 line wavelength. The RIN is measured to be $8 \cdot 10^{-14}\,\text{Hz}^{-1} = -131\,\text{dB/Hz}$ at 500 Hz and $5 \cdot 10^{-14}\,\text{Hz}^{-1} = -133\,\text{dB/Hz}$ at 1 kHz frequencies. Both values are smaller by at least one order of magnitude than the target values stated in Sect. 2.3.4. The VCSEL frequency noise is inversely proportional to the frequency with a slope of nearly $10^{13}\,\text{Hz}^2$ between 10 and 10^5 Hz.

7.2.3 Modulation Sideband Characteristics

The bias current of the VCSEL from Fig. 7.2 is modulated by an RF signal at a frequency that is equal to half of the Cs hyperfine ground splitting frequency (i.e., modulation frequency $f = 4.596\,\text{GHz}$). This creates modulation sidebands separated by 9.192 GHz, as required for the CPT interaction to take place. Figure 7.8 shows the relative strengths of the carrier and the first-order sidebands in dependence of the RF modulation power, measured using a scanning Fabry–Pérot interferometer.

The modulation indices M and R extracted by fitting the sideband powers resulting from (3.44) to the experimental data along with the asymmetry factor S are depicted in Fig. 7.8. As can be seen, $S \approx 2\,\text{dB}$ remains almost constant with applied modulation

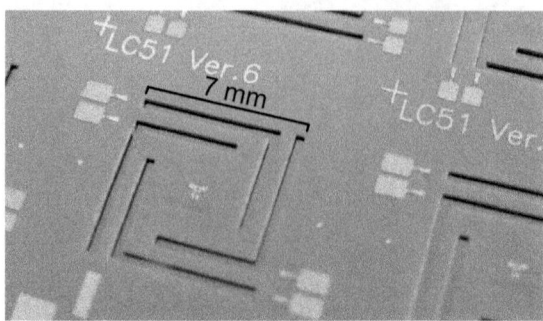

Figure 7.9: Wafer-level LTCC platforms for VCSEL subsystems. The VCSEL chip is bonded to the center of such a platform. A scale has been added to the photograph (from [6]).

power, while R slightly changes from 0.05 to nearly 0.18. Both first-order sidebands become stronger than the carrier at modulation powers of $\geq +5\,\mathrm{dBm}$, and M increases but does not exceed 1.65. According to (3.45), the Henry factor α_H increases from -11 to -8.5 over the applied range of modulation powers. The magnitude of α_H determined by this method is higher than usually expected; see (3.43). The optimum value of M for the CPT interaction is 1.8, at which most of the RF modulation power is transferred to the first-order sidebands [208]. Limitations in modulation efficiency can be attributed to imperfect coupling of the modulation signal caused by impedance mismatch between the RF source and the TO-46 can, which was evidenced by approximately 40% power reflection [206]. However, a CPT signal with a narrow linewidth is obtained, as will be shown in Sect. 7.3. To solve the problem of modulation efficiency limitation for MACs, VCSELs are designed to be flip-chip mounted on LTCC (low-temperature co-fired ceramic) platforms as part of the clock microsystem. The platforms contain laser driver ASICs (application-specific integrated circuits) and impedance matching circuits. The latter would considerably reduce electrical parasitics and make the modulation more efficient, thus minimizing the required RF power. LTCC platforms are displayed in Fig. 7.9 on a wafer before separation. Afterwards the VCSEL, the driver ASIC and the impedance matching circuit are bonded onto the platform.

7.3 CPT Resonance Signal Measurement

As mentioned at the beginning of this chapter, instead of a miniaturized atomic clock, an experimental larger-scale atomic clock demonstrator has been set up at FEMTO-ST. The main advantage of such a demonstrator is the validation of the functionality of the concept

7.3 CPT Resonance Signal Measurement

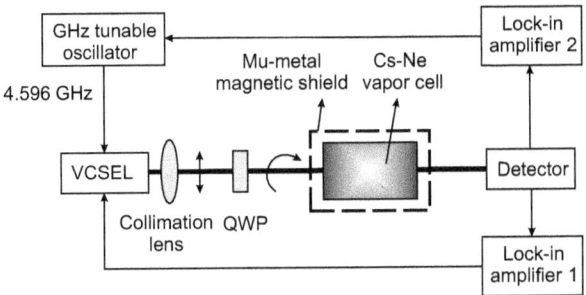

Figure 7.10: Simplified schematic drawing of the experimental atomic clock demonstrator at FEMTO-ST.

of MACs without the hard constraints of miniature packaging and integration. Figure 7.10 shows a simplified schematic of the experimental atomic clock setup. The employed laser source is the grating relief VCSEL from Fig. 7.5 mounted in a TO-46 can, as described in Sect. 7.1.2. It is operated at 1.4 mA and 30°C to reach the Cs D_1 line wavelength, as can be seen from Fig. 7.5 (right). The linewidth of the grating relief VCSEL is measured to be approximately 25 MHz, which is close to the value measured for the standard VCSELs. The optical output of the laser is collimated to a 2 mm diameter beam. The collimated laser beam is then circularly polarized using a quarter-wave plate (QWP) and transmitted through the microfabricated vapor cell. The incident laser power to the cell is about 22 µW. The injection current of the laser is modulated at 4.596 GHz using a commercial frequency synthesizer to produce two modulation sidebands separated by 9.192 GHz, as required for the CPT interaction. The modulation power is set to -2 dBm for a maximum amplitude of the CPT signal. The Cs cell was described in Sect. 2.3.2 and is filled with Ne buffer gas for its favorable properties regarding the short- and long-term frequency stability of the clock as mentioned in Sect. 2.3.2. The cell is stabilized to a temperature of approximately 80°C, at which the sensitivity of the clock frequency to temperature variations is canceled [48, 51]. A static magnetic field of a few µT parallel to the laser beam is applied to split the Zeeman sublevels and enable a σ^+-σ^+ CPT clock scheme on the Cs D_1 line. More information on Zeeman sublevels and the σ^+-σ^+ CPT scheme can be found in App. A.2. To prevent the influence of earth or environmental magnetic field perturbations, the cell is placed in a single-layer mu-metal magnetic shield. The σ^+-σ^+ CPT scheme on the Cs D_1 line employs $|F = 3, m_F = 0\rangle$ and $|F = 4, m_F = 0\rangle$ Zeeman sublevels as the two ground energy levels in the Λ-system, as illustrated in App. A.2. The σ^+ optical transition requires circularly polarized light, hence the necessity of the QWP.

7 Experimental Cesium-Based Atomic Clock Demonstrator

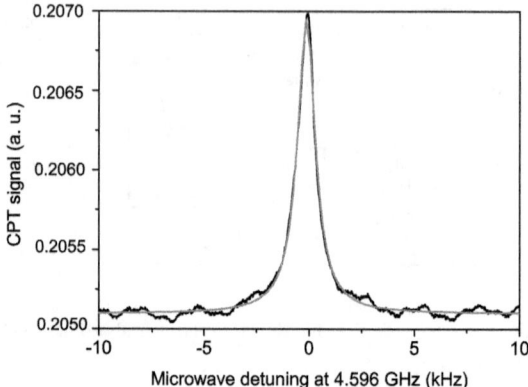

Figure 7.11: Measured CPT resonance signal (black line) for a Cs–Ne microfabricated cell along with a Lorentzian curve fit (red line). The temperature of the cell is 80°C (figure courtesy of FEMTO-ST) [206].

The laser power is detected at the output of the cell with a photodiode that provides signals for two lock-in amplifiers. The first one serves to stabilize the laser emission frequency at the Cs D_1 line, while the second provides the CPT signal, which is used to stabilize the GHz tunable oscillator at half of the Cs hyperfine ground splitting frequency. Figure 7.11 displays a CPT resonance signal obtained at the output of the second lock-in amplifier. The CPT signal is fit by a Lorentzian function showing a linewidth of 1.04 kHz FWHM and a contrast $\bar{C}$ of 0.93% at 80°C cell temperature [206]. The linewidth does not vary significantly in the 70–90°C temperature range. The expected short-term instability of the clock from the CPT signal is $\sigma_{\bar{y}}(\tau) = 2.9 \times 10^{-11}(\tau/\text{s})^{-0.5}$ at $\tau \approx 1\,\text{s}$ averaging times [206]. The result proves that the VCSELs are very good candidates for the development of high-performance Cs-based MACs.

Chapter 8

Conclusion

This dissertation gives a detailed study on design, fabrication, characterization and improvements of vertical-cavity surface-emitting lasers for cesium-based miniaturized atomic clocks. As explained in Chap. 2, MACs employ the all-optical coherent population trapping method. The importance of MACs originates mainly from their high capability to bridge the performance gap between quartz-based oscillators and rubidium atomic frequency standards at a reduced price. The fundamental properties and functionalities that VCSELs possess, discussed in Chap. 3, make them the most suitable light sources for MACs. The main requirements on VCSELs, such as correct emission wavelength, polarization-stable single-mode emission, low threshold currents, high-temperature operation, narrow linewidth, low relative intensity noise, low frequency noise and sufficient modulation bandwidth have been successfully met.

The transfer of an 850 nm VCSEL design to the target wavelength of 894.6 nm has been achieved by introducing indium into the GaAs quantum wells to adjust the emission wavelength and by scaling all layer thicknesses. Detailed quantum-mechanical computations were done in order to evaluate the required indium content, as discussed in Chap. 4. The influence of compressive strain, the band offsets of an InGaAs/AlGaAs heterojunction, as well as the effect of bandgap renormalization have been considered. In fabricated samples, the indium content has been adjusted to achieve a quasi-matching between cavity resonance and optical gain peak at elevated ambient temperature. It has been shown experimentally that an 894.6 nm wavelength VCSEL with quantum wells having 4% indium content exhibits a minimum threshold current at $T = 75°C$ substrate temperature.

As explained in Chap. 4, the inhomogeneity of wafers grown with the employed MBE machine leads to a very small yield of VCSELs with the target emission wavelength of 894.6 nm. Variations of the wet-thermal oxidation rate, which are also due to the inhomogeneity of layer thicknesses, produce a variety of sizes of the active apertures of the devices. This leads to fluctuations of the SMSR and can even cause multi-mode lasing.

8 Conclusion

Hence, a further reduction of suitable VCSELs for MACs occurs. Nevertheless, industry-level epitaxy machines are capable of producing much more homogeneous VCSEL wafers and thus much higher yield compared to a research-level MBE. Still, mainly the tight wavelength specification of atomic clock VCSELs will probably reduce the device yield to levels below that of datacom VCSELs, which finally translates into higher chip cost.

For polarization control, a previously developed technique relying on the integration of a semiconducting surface grating in the top Bragg mirror of the VCSEL structure is employed. More specifically, first of all a so-called inverted grating has been utilized, as presented in Chap. 4, where the grating is etched in an extra topmost GaAs quarter-wave anti-phase layer. As experimentally introduced in Chap. 5, the VCSELs are polarized orthogonal to the grating lines with orthogonal polarization suppression ratios (OPSRs) exceeding 25 dB and a peak-to-peak difference between the dominant and the suppressed polarization modes reaching 40 dB at $T = 80\,°C$. The polarization stability has been investigated at different elevated substrate temperatures up to $T = 80\,°C$, where the VCSEL remains polarization-stable even well above thermal roll-over.

A different type of surface grating which is etched in an extra topmost GaAs half-wave in-phase layer, referred to as regular grating, has been employed, as presented in Chap. 6. Detailed numerical simulations for regular and inverted grating VCSEL designs were presented in Sect. 4.4 and showed that regular grating VCSELs have inherently lower threshold gains than equivalent inverted grating devices. Experimental results of regular grating VCSELs with etch depths of $d \approx \lambda_{mat}/2$, presented in Sect. 6.2.1, show that the threshold currents are reduced to 40% compared to the devices employing inverted gratings. The output polarization is parallel to the grating lines with OPSRs exceeding 20 dB and a peak-to-peak difference between the dominant and the suppressed polarization modes of 25 dB even at $T = 80\,°C$. It is found that VCSELs with regular surface gratings show, on average, lower |OPSR| than inverted grating VCSELs with smaller grating depths d. Although the surface gratings increase the contribution of suppressed polarization modes with increasing d, the relative dichroism increases, as shown by the simulations presented in Sect. 4.4. As future investigations one could fabricate regular grating VCSELs with shallower etch depths (e.g., $d \approx \lambda_{mat}/4$) and characterize their threshold current I_{th} and |OPSR|. According to the simulations presented in Sect. 4.4, the I_{th} of such shallow regular grating VCSELs are expected to be lower compared to the inverted grating devices with the same d. Simultaneously, such shallow regular grating VCSELs are expected to provide higher |OPSR| values comparable to those of the inverted grating VCSELs. Owing to their relatively low |OPSR|, the regular grating devices presented in this dissertation (i.e., $d \approx \lambda_{mat}/2$) are not recommended for MACs despite their superior threshold currents compared to the inverted grating VCSELs.

An alternative approach to a full-area grating is to etch a surface grating over a circular area of only 3 to 4 µm diameter in the center of the outcoupling facet. Such inverted grating relief VCSELs simultaneously provide favorable enhanced fundamental-mode emission as well as polarization-stable laser oscillation. This results in polarization-stable single-mode VCSELs with somewhat larger active diameters, which can provide greater potential for increased device lifetime, as indicated by the preliminary lifetime tests presented in Sect. 6.4 and much desired for MAC applications. In particular, larger active diameter single-mode devices are preferred, since they degrade much slower in terms of optical output power. Hence, the induced light shift effect, which limits the long-term stability of the MAC, is less predominant. As shown in Sect. 6.2.2, VCSELs with 5 µm active diameter show side-mode suppression ratios of 20 dB even at currents close to thermal roll-over with OPSRs better than 20 dB at elevated ambient temperatures up to $T = 100°C$. However, inherently higher optical losses, even of the fundamental mode, lead to slightly larger threshold currents compared to the full-area grating devices. Such a problem could be solved by adding a few mirror pairs to the top DBR. More investigations on this should be done in the future. For all surface grating approaches with sub-emission-wavelength grating periods, there are no side-lobes in the emission far-fields, confirming the absence of diffraction losses in air.

For the purpose of integration with the atomic clock microsystem, flip-chip-bondable VCSEL chips have been realized. Sub-mA threshold currents and sufficient output powers in the milliwatt range are achieved. Small-signal modulation characteristics of VCSELs are presented in Chap. 5. The required modulation bandwidth of more than 5 GHz is always reached close above the sub-mA threshold currents. Therefore, the devices consume very small power. Maximum bandwidths of above 9 GHz have been measured even at elevated temperatures up to $T = 80°C$. Modulation current efficiency factors larger than $7\,\text{GHz}/\sqrt{\text{mA}}$ are achieved even at $T = 80°C$. Moreover, the intrinsic modulation characteristics of the VCSELs are investigated by precise curve fits of the measured small-signal modulation response curves and relative intensity noise spectra. K-factors of less than 0.4 ns and damping-limited maximum 3 dB corner frequencies exceeding 22 GHz and reaching approximately 31 GHz have been obtained even at $T = 65°C$. To determine the electrical parasitic bandwidth of the VCSEL chip, an equivalent-circuit model has been proposed. Its electrical parasitic components account for the structure of the chip and can be determined by measuring the microwave reflection spectra $S_{11}(f)$. Electrical parasitic bandwidths of 5 to 6 GHz have been obtained.

Two revised flip-chip-bondable VCSEL chip designs with smaller mesa sizes and thinner passivation layers, compared to the laser chips presented in Chap. 5, have been implemented, as described in Chap. 6. The VCSELs of both designs show almost similar

8 Conclusion

electrical parasitic bandwidths of 9 to 10 GHz which are larger by 3 to 5 GHz than the parasitic bandwidths found in Chap. 5. One revised chip design has smaller mesa sizes with vertical side walls fabricated by dry-etching instead of chemical wet-etching, which results in inclined side walls, as is the case for the second design. Therefore, the first design has a smaller area of the oxide layer and hence a reduced parasitic capacitance. Moreover, it has a thinner passivation layer which requires just one mesa etching step down to the first layer pairs of the n-type DBR and a single polyimide layer. On the other hand, the second design requires two mesa etching steps with different diameters reaching the highly n-doped contact layer. For planarization and passivation it requires a deposition of three polyimide layers by means of three lithographic steps. Thus, the processing of the first design is much less complex and more desired for large production volumes of atomic clock VCSELs. The dynamic properties have been introduced in Sect. 6.3. The required modulation bandwidth of 5 GHz is reached close above the sub-mA threshold current. Modulation current efficiency factors larger than $6\,\mathrm{GHz}/\sqrt{\mathrm{mA}}$ are achieved even at $T = 80°C$. K-factors of less than 0.4 ns corresponding to a damping-limited maximum 3 dB corner frequency exceeding 22 GHz at $T = 80°C$ are measured.

Some special characteristics of atomic clock VCSELs have been investigated by MAC-TFC project partners, as discussed in Chap. 7. These characteristics are critical in defining the performance of the final MAC prototype and include emission linewidth, relative intensity and frequency noises in the low-frequency range, as well as modulation sideband characteristics. The results confirm the high-level performance of the VCSELs and prove their suitability as laser sources for cesium-based MACs. Emission spectral linewidth, relative intensity noise and frequency noise of about 23 MHz, $< -132\,\mathrm{dB/Hz}$ at 1 kHz and $< 1 \cdot 10^{10}\,\mathrm{Hz^2/Hz}$ at 1 kHz, respectively, have been measured. An experimental large-scale atomic clock demonstrator has been realized employing an inverted grating relief VCSEL with the reduced complexity flip-chip-bondable design mounted in a TO-46 can. As can be concluded from the above discussion, such VCSELs are very suitable for MACs due to their enhanced reliability and thus long-term stability as well as their much reduced processing complexity, which is highly desirable for large production volumes. The demonstrator has been successfully operated, showing a CPT resonance signal with narrow linewidth of approximately 1 kHz FWHM. The result proves that the employed VCSELs are the best candidates for the development of high-performance cesium-based MACs. Unlike MACs, for the large-scale atomic clock demonstrators, sufficient RF modulation power can be injected and could thus compensate power reflections caused by impedance mismatch between the RF source and the VCSEL incorporated in its TO-can housing. Therefore, a much simpler VCSEL chip design with substrate-side n-contacts (i.e., non flip-chip-bondable) could be employed for the large-scale clocks.

Appendix A

Cesium Properties

In this appendix some physical and optical properties of cesium that are relevant to various quantum optical experiments (e.g., cesium-based atomic clocks) are presented. Additionally, some of the parameter concepts which are required for the treatment of interaction between coherent light and cesium atoms are briefly discussed.

A.1 Fine and Hyperfine Structure

Cesium (its chemical symbol: Cs) is an alkali metal which has 55 electrons. Its electron configuration is represented by $1s^2\, 2s^2\, 2p^6\, 3s^2\, 3p^6\, 4s^2\, 3d^{10}\, 4p^6\, 5s^2\, 4d^{10}\, 5p^6\, 6s^1$, where s, p, d are the orbital angular momentum states of the electrons[1]. It has only one valence electron in the outer shell which is identified by a principal quantum number $n_Q = 6$. ^{133}Cs is the only stable isotope and therefore is the one employed in atomic clocks. The following discussion is particularly about this isotope.

For the valence electron, there are two possible excited levels formed by the so-called *fine structure* giving the possibility for two atomic transitions which are known as D_1 and D_2 lines as depicted in Fig. A.1. The fine structure is a result of the interaction between the total orbital angular momentum **L** of the atom and the total spin angular momentum **S**. The total electron angular momentum **J** of the atom can be thus given by the vector sum of **L** and **S** as [209, 210]

$$\mathbf{J} = \mathbf{L} + \mathbf{S}, \tag{A.1}$$

where the bold symbols are vector operators. Each momentum vector has an associated quantum number denoting the possible eigenvalues of its squared magnitude. Thus, the

---
[1]The first number is the principal quantum number n_Q and the superscript is the number of electrons in each orbital state. For the orbital angular momentum states of the atom as a whole, a capital letter notation, i.e., S, P, D, is rather used [209].

A Cesium Properties

vectors **J**, **L** and **S** have the quantum numbers J, L and S, respectively, by which the squared magnitudes [209]

$$|\mathbf{L}|^2 = \mathbf{L} \cdot \mathbf{L} = L(L+1)\hbar^2, \tag{A.2a}$$

$$|\mathbf{S}|^2 = \mathbf{S} \cdot \mathbf{S} = S(S+1)\hbar^2, \tag{A.2b}$$

$$|\mathbf{J}|^2 = \mathbf{J} \cdot \mathbf{J} = J(J+1)\hbar^2 \tag{A.2c}$$

are defined. The vector operators can additionally have quantum numbers which denote the possible eigenvalues of one of its components (usually the one which is along the atomic quantization axis z). The z components

$$L_z = m_L \hbar, \tag{A.3a}$$

$$S_z = m_S \hbar, \tag{A.3b}$$

$$J_z = m_J \hbar \tag{A.3c}$$

are calculated using the quantum numbers m_L, m_S and m_J, respectively. From the addition rules of the angular momentum vectors, J has to be within the range [209, 210]

$$|L - S| \leq J \leq L + S, \tag{A.4}$$

and the quantum numbers

$$m_L = -L, -L+1, \cdots, L-1, L, \tag{A.5a}$$

$$m_S = -S, -S+1, \cdots, S-1, S, \tag{A.5b}$$

$$m_J = -J, -J+1, \cdots, J-1, J \tag{A.5c}$$

can be also defined. Energy levels are split according to their J value. The ground level in a Cs atom with $L = 0$ and $S = 1/2$ has a single value of $J = 1/2$ and shows no fine splitting. The excited level with $L = 1$ and $S = 1/2$ splits into one level with $J = 1/2$ and one with $J = 3/2$. Consequently, the D transition line is split into two components, i.e., D_1 and D_2 corresponding to the transitions $6^2S_{1/2} \rightarrow 6^2P_{1/2}$ and $6^2S_{1/2} \rightarrow 6^2P_{3/2}$, respectively. The labeling of the energy levels is based on the so-called *Russell–Saunders notation*. The first number is the principal quantum number n_Q of the valence electron, the superscript is $2S + 1$, the letter refers to L (i.e., for S, $L = 0$, for P, $L = 1$, for D, $L = 2$, etc.), and the subscript gives the value of J.

The *hyperfine structure* of the energy levels arises from the interaction of the total nuclear spin angular momentum **I** and the total electron angular momentum **J**. The total atomic angular momentum **F** is given by [210, 211]

$$\mathbf{F} = \mathbf{J} + \mathbf{I}. \tag{A.6}$$

A.1 Fine and Hyperfine Structure

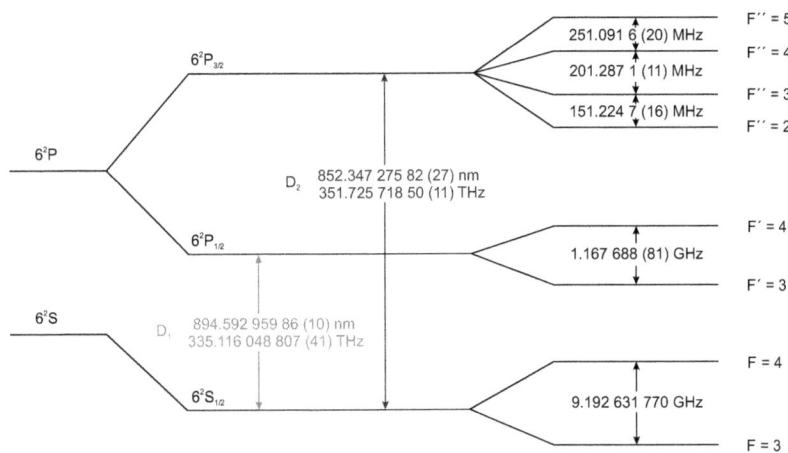

Figure A.1: Cs energy level diagram showing the fine and hyperfine structure of the ground level (6^2S) and the first excited level (6^2P) along with the corresponding D_1 and D_2 transition lines. The frequency splittings between the hyperfine energy levels are taken from [212] and [213] for the excited $6^2P_{1/2}$ and $6^2P_{3/2}$ levels, respectively, while the hyperfine splitting of the ground levels is according to the present SI definition of the second [2]. The values for the D_1 and D_2 transitions are taken from [212] and [214], respectively.

Similar to (A.4), the quantum number F can take the values [210]

$$|J - I| \leq F \leq J + I, \qquad (A.7)$$

where I is the quantum number associated with **I**, and similarly to (A.5) the quantum number m_F can take the values

$$m_F = -F, -F+1, \cdots, F-1, F. \qquad (A.8)$$

To distinguish between the hyperfine structures of different fine energy levels, F' and F'' are assigned to the excited levels $6^2P_{1/2}$ and $6^2P_{3/2}$ of the D_1 and D_2 transitions, respectively, while the hyperfine structure of the ground level $6^2S_{1/2}$ is identified by F, as depicted in Fig. A.1. For $6^2S_{1/2}$, $J = 1/2$ and $I = 7/2$, so $F = 3$ or 4. For $6^2P_{1/2}$, F' is either 3 or 4 and for $6^2P_{3/2}$, F'' can take any of the values 2, 3, 4, or 5. Similarly to the fine structure, the atomic energy levels of the hyperfine structure are shifted according to the value of F, as depicted in Fig. A.1, where the frequency shifts between different fine and hyperfine energy levels are taken from [2, 212–214].

A Cesium Properties

A.2 Zeeman Splitting

Each hyperfine level consists of $2F+1$ magnetic sublevels denoted by the quantum number m_F, which is an integer ranging from $-F$ to $+F$, as defined in (A.8). In the absence of an external magnetic field, these sublevels are degenerate, i.e., they have the same amount of energy. However, when an external static magnetic field is applied, their degeneracy is lifted and each has an energy according to its m_F value. This effect is called the *Zeeman effect* and the generated sublevels are called *Zeeman sublevels*. The external static magnetic field has to be weak (i.e., in the range of a few µT), so the magnetically-induced energy shifts are small compared to the hyperfine splittings. Otherwise, the hyperfine structure (e.g., required for the atomic clock application) would be strongly perturbed. As an example, the hyperfine energy level $F' = 4$ of the excited level $6^2P_{1/2}$ can be split by an external static magnetic field into 9 Zeeman sublevels, each of which is denoted by $m_{F'}$ ranging from -4 to 4, as illustrated in Fig. A.2. In total, the four hyperfine levels are split into 32 Zeeman sublevels, as can be seen from the figure.

A concise and convenient way to describe the quantum states of an atom is to use *Dirac's notation* [215]. For instance, a Zeeman sublevel in the ground level $6^2S_{1/2}$ having the quantum numbers $F = 4$ and $m_F = 0$, one can say that at this level, the atom has a *quantum state* of $|F = 4, m_F = 0\rangle$, where the notation $|\rangle$ is called *Dirac's ket*. This notation would be helpful for describing the optical pumping and its associated atomic transitions for Cs in what follows.

So far, Cs atoms are considered to be isolated or subjected to an external static magnetic field. However, for the atomic clock operation, Cs atoms are additionally subjected to laser emission. This is called *optical pumping* and the description should be thus expanded to the interaction of the atoms with coherent light. Atomic transitions for Cs atoms subjected to a weak external static magnetic field (i.e., Zeeman splitting of the hyperfine levels occurs, as seen in Fig. A.2) are governed by the following *selection rules* [216]

$$\Delta J = 0, \pm 1, \tag{A.9a}$$
$$\Delta L = 0, \pm 1, \tag{A.9b}$$
$$\Delta S = 0, \tag{A.9c}$$
$$\Delta F = 0, \pm 1, \tag{A.9d}$$
$$\Delta m_F = 0, \pm 1. \tag{A.9e}$$

For the D_1 line $\Delta J = 0$, $\Delta L = 1$ and $\Delta S = 0$, while for the D_2 line, they are the same except for ΔJ which has a value of 1. For both D lines, ΔF and Δm_F are according to (A.9d) and (A.9e), respectively.

A.2 Zeeman Splitting

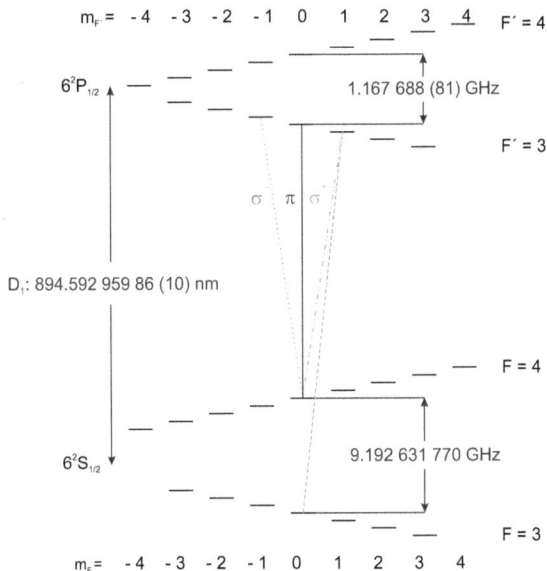

Figure A.2: Zeeman sublevels in the hyperfine structure of the Cs D_1 line. The possible atomic transitions from the ground state $|F = 4, m_F = 0\rangle$ to the excited states $|F' = 3, m_{F'} = 1\rangle$ (dash-dotted green), $|F' = 3, m_{F'} = 0\rangle$ (solid blue) or $|F' = 3, m_{F'} = -1\rangle$ (dotted red) are shown and assigned the symbols σ^+, π and σ^- which denote atomic transitions using right-handed circularly, linearly or left-handed circularly polarized light radiation, respectively. The $\sigma^+ - \sigma^+$ CPT scheme is indicated by the dashed and dash-dotted green lines.

For the following discussion, only the D_1 line and in particular the $|F = 4, m_F\rangle \rightarrow |F' = 3, m_{F'}\rangle$ transitions are considered. Depending on the polarization of the light radiation, one of the three values of Δm_F from (A.9e) applies. For right-handed circularly polarized light with its propagation direction aligned to the external static magnetic field, the $|F = 4, m_F\rangle \rightarrow |F' = 3, m_{F'} = m_F + 1\rangle$ transition occurs with $\Delta m_F = m_{F'} - m_F = 1$. This atomic transition is known as σ^+ transition. For left-handed circularly polarized light, the $|F = 4, m_F\rangle \rightarrow |F' = 3, m_{F'} = m_F - 1\rangle$ transition occurs with $\Delta m_F = -1$. This transition is known as σ^- transition. Finally, for linearly polarized light with its propagation parallel to the external static magnetic field, the $|F = 4, m_F\rangle \rightarrow |F' = 3, m_{F'} = m_F\rangle$ transition occurs with $\Delta m_F = 0$, where such a transition is known as π transition. The three atomic transitions are displayed in Fig. A.2, where $|F = 4, m_F = 0\rangle$ is considered as the initial quantum state of the Cs atom.

A Cesium Properties

The selected scheme for the atomic clock presented in this dissertation is $\sigma^+\text{-}\sigma^+$ CPT on the Cs D_1 line, for which the $|F=3, m_F=0\rangle$ and $|F=4, m_F=0\rangle$ Zeeman sublevels are employed as the two ground quantum states in the Λ-system from Fig. 2.1 (a). Since the two $m_F = 0$ ground states are participating, it is also known as the *0–0 transition* scheme. For weak environmental magnetic field perturbations, the energies of these two Zeeman sublevels have a small quadratic dependence on the field [2]. Therefore, the magnetic shielding required for the Cs vapor cell is more tolerant compared to cases where Zeeman sublevels with $m_F \neq 0$ (whose energies show a larger, linear dependence on environmental magnetic field perturbations) are employed. The associated atomic transitions for the $\sigma^+\text{-}\sigma^+$ CPT scheme are $|F=3, m_F=0\rangle \rightarrow |F'=3, m_{F'}=1\rangle$ and $|F=4, m_F=0\rangle \rightarrow |F'=3, m_{F'}=1\rangle$, which are displayed in Fig. A.2 by dashed and dash-dotted green lines, respectively.

Appendix B

MAC-TFC Consortium

The MAC-TFC consortium consists of 10 project partners gathering all the required competences to develop and demonstrate necessary technology to achieve miniaturized, low-power, atomic frequency reference units. The consortium constitutes of a multi-disciplinary research team involving academic partners, international research institutes, and industrial enterprises, namely [6]:

No.	Abbreviation	Name	Location
1.	FEMTO-ST	Franche-Comté Electronique Mécanique Thermique et Optique Sciences et Technologies	Besançon, France
2.	UniNE-LTF	Université de Neuchâtel-Laboratoire Temps-Fréquence	Neuchâtel, Switzerland
3.	EPFL	Ecole Polytechnique Fédérale de Lausanne	Neuchâtel, Switzerland
4.	PWR	Politechnika Wroclawska	Wrocław, Poland
5.	UULM	Ulm University	Ulm, Germany
6.	VTT	VTT Technical Research Centre of Finland	Oulu, Finland
7.	CEA	Commissariat à l'Énergie Atomique et aux Energies Alternatives	Grenoble, France
8.	SAES	SAES Getters S.p.A.	Milan, Italy
9.	OSA	Oscilloquartz SA-Swatch Group	Neuchâtel, Switzerland
10.	SWATCH R&D	The Swatch Group Recherche et Développement SA	Neuchâtel, Switzerland

Appendix C

Mask Layouts

Figure C.1 depicts a schematic layout of a small part of the lithographic mask designed to process flip-chip-bondable VCSELs for MACs. In particular, the figure shows the layout for a solitary VCSEL device. The mask contains 10 × 10 unit cells, where each cell is made of a 10 × 10 VCSEL array with 300 µm pitch and 0.5 µm step increase in the p-mesa size per row[1].

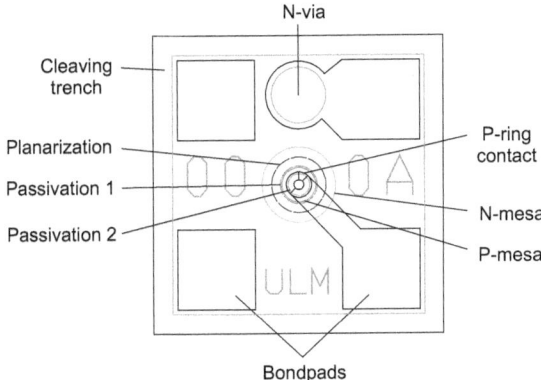

Figure C.1: Schematic layout of the lithographic mask used to perform the processing steps needed to fabricate flip-chip-bondable VCSELs for MACs.

---

[1]Each row of VCSELs has thus similar oxide aperture diameters. Therefore, each unit cell has 10 different oxide diameters with a maximum difference of 4.5 µm between the largest and the smallest devices.

C Mask Layouts

Appendix D

VCSEL Processing

In what follows, detailed processing steps for flip-chip-bondable VCSELs with surface grating are described for thick and thin planarization technologies. Both processes are based on the mask design presented in App. C. Each lithographic step is preceded by sample cleaning in acetone and isopropanol and drying on a hot-plate at 120°C for 10 min unless otherwise stated. For optical exposure, a mask aligner[1] with an emission wavelength of 320 nm is used. For the electron beam exposure, an electron beam lithography system[2] is utilized.

D.1 Flip-Chip-Bondable VCSELs with Thick Planarization Layers

1. Surface grating

- Drying: hot-plate, 180°C, 15 min
- Spin coating: PMMA 950K 4:1[3], 2500 rpm, 60 s
- Prebake: hot-plate, 180°C, 5 min
- Exposure: 470 µC/cm^2 dose, 50 kV acceleration voltage
- Developing: methyl isobutyl ketone : isopropanol = 1:3, 90 s
- Postbake: hot-plate, 110°C, 3 min
- O$_2$-plasma: 10% O$_2$, 100 W RF, 2 min
- HCl-dip: HCl:H$_2$O = 1:1, 15 s
- Etching: preparation: citric acid : H$_2$O = 1:1, then add H$_2$O$_2$ 1:30, etch rate ≈ 1.1 nm/s

---

[1]Karl Suss, model MJB3.
[2]Leica E-Beam Lithography Systems, model EBPG 5 HR.
[3]polymethyl methacrylate (PMMA) with a molecular weight of 950K solved in ethyl lactate 4:1.

D VCSEL Processing

- Resist removal: methylpyrrolidone, 100°C, 5 min
- Cleaning: acetone, isopropanol

2. P-mesa etching and oxidation
Wet-etched mesa

- Spin coating: AZ1512 HS, 6000 rpm, 40 s, use AZ-EBR to remove the bead from the edges
- Prebake: hot-plate, 90°C, 10 min
- Exposure: 12 mW/cm^2, 14 s, mask: P-MESA-B
- Developing: AZ400K:H$_2$O = 1:4, 18 − 20 s
- Postbake: hot-plate, 120°C, 2 min
- Etching: H$_2$SO$_4$ (98%):H$_2$O$_2$:H$_2$O = 1:6:40, etch rate ≈ 1.2 μm/s, stop after etching the first mirror pair of the n-mirror
- Resist removal: methylpyrrolidone, 100°C, 5 min
- Cleaning: acetone, isopropanol
- Oxidation: furnace 370°C, bubbler 96°C, 0.5 l/min N$_2$, 17.5 min

Dry-etched mesa

- Spin coating: AZ4533, 6000 rpm, 40 s, use AZ-EBR to remove the bead from the edges
- Prebake: hot-plate, 90°C, 15 min
- Exposure: 12 mW/cm^2, 120 s, mask: P-MESA-C
- Developing: AZ400K:H$_2$O = 1:4, 60 s
- Postbake: hot-plate, 100°C, 2 min
- Etching: RIE: 10 sccm SiCl$_4$, 6 sccm Ar, 35 W, 9 mT, 50 min
- O$_2$-plasma: 20% O$_2$, 10% CF$_4$, 100 W RF, 3 min
- Resist removal: methylpyrrolidone, 100°C, 30 min
- Cleaning: acetone, isopropanol
- Oxidation: furnace 370°C, bubbler 96°C, 0.5 l/min N$_2$, 14 min

3. N-mesa etching and n-contact

- Spin coating: AZ4533, 6000 rpm, 40 s, use AZ-EBR to remove the bead from the edges
- Prebake: hot-plate, 90°C, 15 min
- Exposure: 12 mW/cm^2, 120 s, mask: N-MESA-B
- Developing: AZ400K:H$_2$O = 1:4, 50 s
- Postbake: hot-plate, 100°C, 2 min

D.1 Flip-Chip-Bondable VCSELs with Thick Planarization Layers

- HF-dip: buffered hydrofluoric acid (NH_4F:HF = 87.5:12.5), 15 s
- Etching: H_2SO_4 (98%):H_2O_2:H_2O = 1:6:40,
  etch rate ≈ 1.2 µm/s, stop after etching the last
  n-mirror layer by 40 s
- Evaporation: Ge (17 nm) − Au (50 nm) − Ni (10 nm) − Au (50 nm)
- Lift-off: methylpyrrolidone, 100°C, 30 min
- Cleaning: acetone, isopropanol

4. P-ring contact

- Spin coating: AZnLOF 2070, 4000 rpm, 40 s, use AZ-EBR to remove
  the bead from the edges
- Prebake: hot-plate, 110°C, 2 min
- Exposure: 12 mW/cm², 8 s, mask: P-CONTACT-B for wet-etched
  mesa or P-CONTACT-C for dry-etched mesa
- Postbake: hot-plate, 110°C, 90 s
- Developing: AZ826MIF, 130 s
- O_2-plasma: 10% O_2, 100 W RF, 2 min
- Evaporation: Ti (20 nm) − Pt (50 nm) − Au (150 nm)
- Lift-off: methylpyrrolidone, 100°C, 30 min
- Cleaning: acetone, isopropanol

5. Polyimide planarization

- Spin coating: Durimide 7520, 8000 rpm, 40 s, use HTR-D2 to remove
  the bead from the edges
- Prebake: hot-plate, 100°C, 5 min
- Exposure: 12 mW/cm², 11 s, mask: PLANARIZATION-B,
  leave a corner for electroplating
- Postbake: hot-plate, 100°C, 1 min
- Developing: HTR-D2, 115 s, stop with isopropanol
- Hard bake: ramp up to 300°C in 1 h, hold for 1 h, cool down

6. Polyimide passivation 1

- Spin coating: Durimide 7520, 8000 rpm, 40 s, use HTR-D2 to remove
  the bead from the edges
- Prebake: hot-plate, 100°C, 5 min
- Exposure: 12 mW/cm², 25 s, mask: PASSIVATION1-B,
  leave a corner for electroplating

- Postbake: hot-plate, 100°C, 1 min
- Developing: HTR-D2, 135 s, stop with isopropanol
- Hard bake: ramp up to 300°C in 1 h, hold for 1 h, cool down

7. Polyimide passivation 2
- Spin coating: Durimide 7505, 9000 rpm, 40 s, use HTR-D2 to remove the bead from the edges
- Prebake: hot-plate, 100°C, 2 min
- Exposure: 12 mW/cm^2, 20 s, mask: PASSIVATION2-B, leave a corner for electroplating
- Postbake: hot-plate, 100°C, 1 min
- Developing: HTR-D2, 150 s, stop with isopropanol
- Hard bake: ramp up to 300°C (or 350°C for wire-bondable bondpads) in 1 h, hold for 1 h, cool down

8. N-via electroplating
- Spin coating: AZ4533, 4000 rpm, 40 s, use AZ-EBR to remove the bead from the edges
- Prebake: hot-plate, 90°C, 15 min
- Exposure: 12 mW/cm^2, 4 min, mask: GALVANIK-B, leave a corner for electroplating
- Developing: AZ400K:H$_2$O = 1:4, 120 s
- O$_2$-plasma: 10% O$_2$, 100 W RF, 2 min
- Postbake: hot-plate, 100°C, 1 min
- Electroplating:
  - cover the back side of the sample by a blue tape
  - switch on the bath circulating pump
  - set the voltage to 0.4 V
  - set pH to 8 using NaOH:H$_2$O = 3:10 or H$_2$O to increase or decrease the pH, respectively
  - leave the bath running for 30 min before use
  - start electroplating and measure the thickness of electroplated gold using a profilometer[4]
  - stop electroplating when gold fills the n-via and becomes higher than polyimide layer by 2 μm
  - remove the blue tape
- Resist removal: methylpyrrolidone, 100°C, 30 min
- Cleaning: acetone, isopropanol

---

[4]Tencor, model Alpha Step 100.

9. Bondpads

- Spin coating: AZnLOF 2070, 9000 rpm, 40 s, use AZ-EBR to remove the bead from the edges
- Prebake: hot-plate, 110°C, 2 min
- Exposure: 12 mW/cm^2, 6 s, mask: BONDPAD-B
- Postbake: hot-plate, 110°C, 90 s
- Developing: AZ826MIF, 115 s
- O_2-plasma: 10% O_2, 100 W RF, 2 min
- Evaporation: with rotation and angle of 70°
  Ni (100 nm) − Au (50 nm) − Ni (100 nm) − Au (100 nm) − Ti (20 nm) − Pt (50 nm) − Au (150 nm)
- Lift-off: methylpyrrolidone, 100°C, 30 min
- Cleaning: acetone, isopropanol
- Annealing[5]: ramp up to 330°C in 1 h, hold for 1 h, cool down

10. Substrate thinning

- Preparation: glue the sample to a glass carrier using Crystalbond[6]
- Thinning: H_2O_2:NH_4, add H_2O_2 until pH = 8.5, etch rate ≈ 6.5 µm/s, stop when substrate thickness ≈ 180 µm

D.2 Flip-Chip-Bondable VCSELs with Thin Planarization Layer (Simpler Processing)

1. Surface grating

As D.1.1

2. P-mesa etching and oxidation

As D.1.2, dry-etched mesa

3. N-contact

- Spin coating: AZ4533, 6000 rpm, 40 s, use AZ-EBR to remove the bead from the edges
- Prebake: hot-plate, 90°C, 15 min
- Exposure: 12 mW/cm^2, 120 s, mask: N-MESA-B
- Developing: AZ400K:H_2O = 1:4, 50 s
- Postbake: hot-plate, 100°C, 2 min

---

[5] Optional step required to achieve wire-bondable bondpads.
[6] T-E-Klebetechnik, type 509.

D VCSEL Processing

- HF-dip: buffered hydrofluoric acid (NH_4F:HF = 87.5:12.5), 15 s
- Evaporation: Ge (17 nm) − Au (50 nm) − Ni (10 nm) − Au (50 nm)
- Lift-off: methylpyrrolidone, 100°C, 30 min
- Cleaning: acetone, isopropanol

4. P-ring contact

- Spin coating: AZnLOF 2070, 9000 rpm, 40 s, use AZ-EBR to remove the bead from the edges
- Prebake: hot-plate, 110°C, 2 min
- Exposure: 12 mW/cm^2, 6 s, mask: P-CONTACT-B for wet-etched mesa or P-CONTACT-C for dry-etched mesa
- Postbake: hot-plate, 110°C, 90 s
- Developing: AZ826MIF, 110 s
- O_2-plasma: 10% O_2, 100 W RF, 2 min
- Evaporation: Ti (20 nm) − Pt (50 nm) − Au (150 nm)
- Lift-off: methylpyrrolidone, 100°C, 30 min
- Cleaning: acetone, isopropanol

5. Polyimide passivation

- Spin coating: Durimide 7505, 4000 rpm, 40 s, use HTR-D2 to remove the bead from the edges
- Prebake: hot-plate, 100°C, 2 min
- Exposure: 12 mW/cm^2, 15 s, mask: PLANARIZATION-B, leave a corner for electroplating
- Postbake: hot-plate, 100°C, 1 min
- Developing: HTR-D2, 150 s, stop with isopropanol
- Hard bake: ramp up to 350°C in 1 h, hold for 1 h, cool down

6. N-via electroplating

As D.1.8

7. Bondpads

As D.1.9

8. Substrate thinning

As D.1.10

Appendix E

VCSEL Epitaxial Structure

The following table provides the VCSEL epitaxial layer structure of this dissertation grown using solid-source MBE. The layers are specified by thickness and material composition along with their function in the device. The thickness of the cap layer t_{cap} is set to be $\lambda_{\text{mat}}/4$ for an inverse VCSEL structure or $\lambda_{\text{mat}}/2$ for a regular structure. For different samples, the number of layer pairs N_{p} of the upper p-type DBR is varied to be either 25 or 28. The active region contains three compressively strained $\text{In}_x\text{Ga}_{1-x}\text{As}$ QWs with $x = 6\%$, 4.5% or 4% indium content.

Repetition	Material	Thickness (nm)	Function
1	GaAs	10.6	cap layer
	GaAs	t_{cap}	
	GaAs	6.7	
N_{p}	$\text{Al}_{0.20}\text{Ga}_{0.80}\text{As}$	31.9	p-type DBR
	$\text{Al}_{0.27}\text{Ga}_{0.73}\text{As} \rightarrow \text{Al}_{0.47}\text{Ga}_{0.53}\text{As}$	13.0	
	$\text{Al}_{0.47}\text{Ga}_{0.53}\text{As} \rightarrow \text{Al}_{0.90}\text{Ga}_{0.10}\text{As}$	20.1	
	$\text{Al}_{0.90}\text{Ga}_{0.10}\text{As}$	31.1	
	$\text{Al}_{0.90}\text{Ga}_{0.10}\text{As} \rightarrow \text{Al}_{0.47}\text{Ga}_{0.53}\text{As}$	20.1	
	$\text{Al}_{0.47}\text{Ga}_{0.53}\text{As} \rightarrow \text{Al}_{0.27}\text{Ga}_{0.73}\text{As}$	13.0	
	$\text{Al}_{0.20}\text{Ga}_{0.80}\text{As}$	39.2	
1	$\text{Al}_{0.27}\text{Ga}_{0.73}\text{As} \rightarrow \text{Al}_{0.47}\text{Ga}_{0.53}\text{As}$	13.0	
	$\text{Al}_{0.47}\text{Ga}_{0.53}\text{As} \rightarrow \text{Al}_{0.90}\text{Ga}_{0.10}\text{As}$	20.1	
	$\text{Al}_{0.90}\text{Ga}_{0.10}\text{As}$	31.1	
	$\text{Al}_{0.90}\text{Ga}_{0.10}\text{As} \rightarrow \text{Al}_{0.47}\text{Ga}_{0.53}\text{As}$	20.1	
	$\text{Al}_{0.47}\text{Ga}_{0.53}\text{As}$	32.6	
	$\text{Al}_{0.47}\text{Ga}_{0.53}\text{As} \rightarrow \text{Al}_{0.90}\text{Ga}_{0.10}\text{As}$	20.1	

E VCSEL Epitaxial Structure

Repetition	Material	Thickness (nm)	Function
1	AlAs	33.3	oxidation layer
	$Al_{0.90}Ga_{0.10}As \rightarrow Al_{0.47}Ga_{0.53}As$	20.1	inner cavity
	$Al_{0.47}Ga_{0.53}As$	85.3	
	$Al_{0.47}Ga_{0.53}As \rightarrow Al_{0.27}Ga_{0.73}As$	43.3	
3	$Al_{0.27}Ga_{0.73}As$	9.9	
	$In_xGa_{1-x}As$	7.9	
	$Al_{0.27}Ga_{0.73}As$	9.9	
	$Al_{0.27}Ga_{0.73}As \rightarrow Al_{0.47}Ga_{0.53}As$	41.5	
	$Al_{0.47}Ga_{0.53}As$	57.8	
1	$Al_{0.47}Ga_{0.53}As \rightarrow Al_{0.90}Ga_{0.10}As$	20.1	n-type DBR
	$Al_{0.90}Ga_{0.10}As$	31.1	
	$Al_{0.90}Ga_{0.10}As \rightarrow Al_{0.47}Ga_{0.53}As$	20.1	
	$Al_{0.47}Ga_{0.53}As \rightarrow Al_{0.27}Ga_{0.73}As$	13.0	
38	$Al_{0.20}Ga_{0.80}As$	39.2	
	$Al_{0.27}Ga_{0.73}As \rightarrow Al_{0.47}Ga_{0.53}As$	13.0	
	$Al_{0.47}Ga_{0.53}As \rightarrow Al_{0.90}Ga_{0.10}As$	20.1	
	$Al_{0.90}Ga_{0.10}As$	31.1	
	$Al_{0.90}Ga_{0.10}As \rightarrow Al_{0.47}Ga_{0.53}As$	20.1	
	$Al_{0.47}Ga_{0.53}As \rightarrow Al_{0.27}Ga_{0.73}As$	13.0	
1	$Al_{0.20}Ga_{0.80}As$	51.9	
	GaAs	2400	n-contact layer

Appendix F

Experimental Measurement Setups

This appendix is dedicated to the measurements setups used to characterize the VCSELs studied during this dissertation.

F.1 Polarization-Resolved Operation Characteristics and Emission Spectra

Figure F.1 shows the experimental setup used to measure the polarization-resolved LIV characteristics and emission spectra of VCSELs presented in this dissertation. This measurement setup was constructed during a previous research work in the VCSEL group [200]. All setup components (indicated by vertical two-sided arrows in Fig. F.1) except the VCSEL holder, the collimating lens and the unit for coupling the light into an optical fiber can be moved in and out of the optical path. The sample is mounted on a vacuum chuck made of copper. The VCSEL temperature is set using a temperature controller unit[1]. The temperature can be increased up to 120°C. The lower temperature limit of the holder is given by the humidity in the laboratory, since at temperatures below approximately 10°C, water vapor condenses on the wafer. As depicted in Fig. F.2, the VCSEL contacts with an SG configuration are accessible from the top by a coplanar microwave probe[2]. Therefore, measurements of the devices can be performed on-wafer without the need for chip separation. The VCSEL is driven by a constant current generated by a laser driver[3]. To measure the optical output power of the VCSEL, its emission is collimated by a lens and directed to a large-area silicon photodetector[4] where a Glan–

---
[1]ILX Lightwave, model LDT-5910B.
[2]Cascade Microtech Inc., model ACP40-SG-200, 40 GHz cut-off frequency.
[3]ILX Lightwave, model LDC-3724B.
[4]Newport Corp., model 818-ST.

F Experimental Measurement Setups

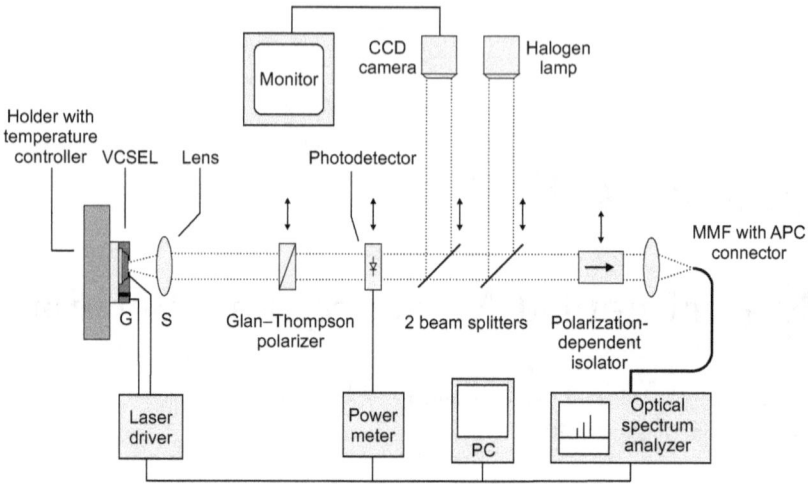

Figure F.1: Schematic drawing of the experimental setup used for measurements of the polarization-resolved LIV curves and emission spectra.

Figure F.2: Photograph of an on-wafer tested VCSEL chip. The device is contacted with a signal–ground microwave probe for operation.

Thompson polarizer[5] can be inserted in front of the photodetector to measure the power in a specific polarization. The laser driver and the power meter[6] of the photodetector are addressed by general purpose interface bus (GPIB), so current and voltage and optical output power can be set or read by a PC. The halogen lamp, the beam splitters and the CCD camera are used to get an enlarged image of the VCSEL device on a monitor, so

---

[5]B. Halle GmbH, model PGT, long version, 60 dB extinction ratio.
[6]Newport Corp., model 1830-C.

F.2 Far-Field Measurements

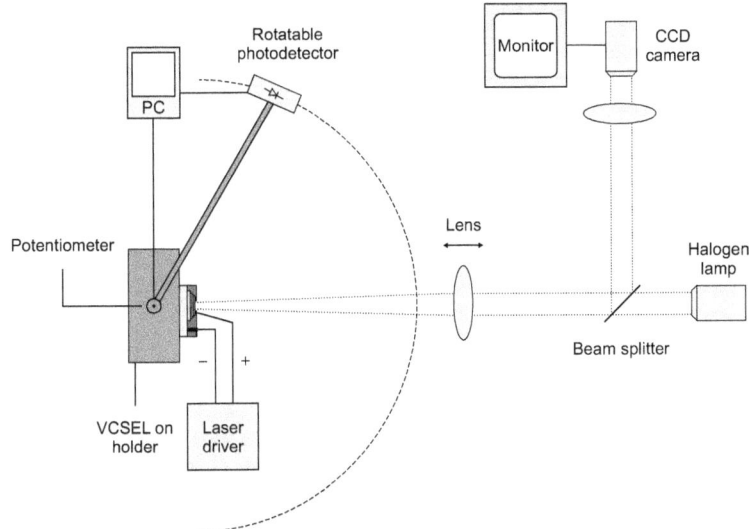

Figure F.3: Schematic drawing of the experimental setup used for measurements of the far-fields.

it can be easily contacted and identified by its label. For spectra measurements, laser light is focused by a lens and injected into a 50 µm core diameter graded-index (GI) MMF. The fiber is connected to an optical spectrum analyzer[7]. A polarization-dependent isolator[8] is utilized as polarizer before the light is coupled into the optical fiber in order to measure polarization-resolved spectra. To avoid unwanted optical feedback, both sides of the MMF are angle-polished[9]. Moreover, anti-reflection-coated lenses are employed and the photodetector is tilted by 5° with respect to the optical axis.

F.2 Far-Field Measurements

The available setup used for far-field measurements is depicted in Fig. F.3. The sample is fixed on a copper mount by vacuum and contacted using two probe needles. A silicon photodetector is attached to a rotatable metal arm in a way that it can be moved along a semi-circle with a radius of approximately 15 cm. The contact probes shadow some

---
[7] ANDO Electric Corp., model AQ 6317, 0.05 nm spectral resolution.
[8] Gsänger Optoelektronik GmbH, model DLI 1, 60 dB isolation.
[9] Angled physical contact (APC) connector: the normal to the front surface of the optical fiber is tilted by 8 degrees with respect to the fiber axis.

F Experimental Measurement Setups

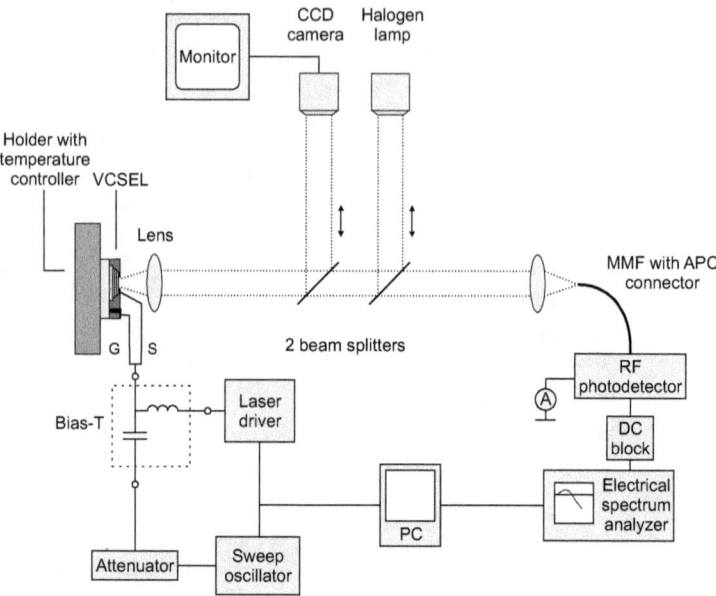

Figure F.4: Schematic drawing of the experimental setup used for measurements of the small-signal modulation response curves.

parts of the semi-circle, thus the photodetector cannot be moved along an entire half-circle. To increase the measurement resolution, the effective area of the photodetector is decreased with a slit aperture of approximately 1 mm. To determine the position angle of the photodetector, a potentiometer is connected to the axis of the metal arm. Finally, an analog-to-digital card is employed to transfer the signals of the photodetector and potentiometer to a PC. Sample illumination and identification of individual devices are done equivalent to Fig. F.1, except that the lens has to be moved away after probing the device in order to characterize the VCSEL without any external optics.

F.3 Small-Signal Modulation Response Measurements

Figure F.4 shows the experimental setup employed to measure the small-signal modulation response curves. The VCSEL is driven by a constant current generated by a laser driver[10] and a low-power RF modulating signal of $-20\,\text{dBm}$ level generated by a sweep

---
[10]ILX Lightwave, model LDC-3724B.

F.3 Small-Signal Modulation Response Measurements

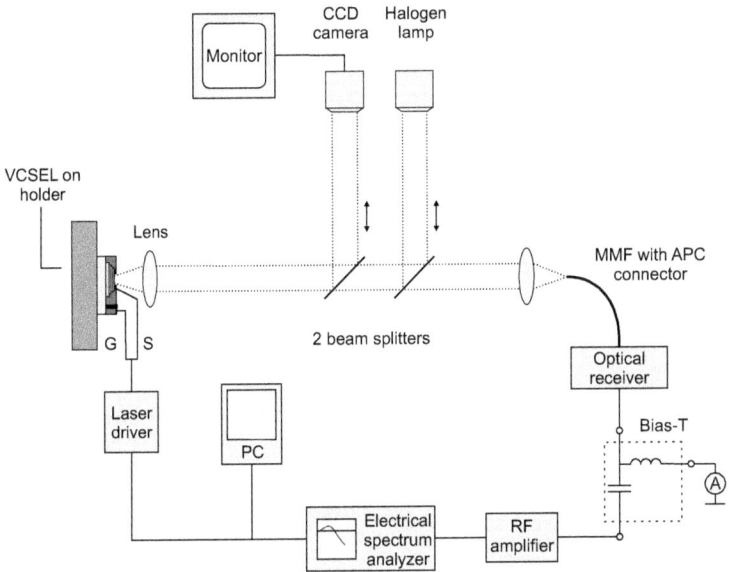

Figure F.5: Schematic drawing of the experimental setup used for measurements of the RIN spectra.

oscillator[11]. The VCSEL chip is contacted via a coplanar microwave probe[12] in the way shown in Fig. F.2. A bias-tee[13] is used to combine the RF and DC current signals. An RF attenuator[14] is directly placed before the bias-tee to attenuate the backward microwave reflections due to impedance mismatch between the VCSEL and the 50 Ω measurement system. The light is coupled via an MMF to an RF photodetector[15] connected to a microwave spectrum analyzer[16] to record the modulation response spectrum. A DC block[17] is placed right at the input of the spectrum analyzer to allow only the RF signal to pass. The frequency characteristics of all cables, connectors, the bias-tee, the attenuator, the DC block, and the RF photodetector are numerically subtracted to obtain only the modulation response of the VCSEL. To avoid unwanted optical feedback, both sides of the MMF are angle-polished. Moreover anti-reflection-coated lenses are employed. All

---
[11]Hewlett Packard, model HP83620A.
[12]Cascade Microtech Inc., model ACP40-SG-200, 40 GHz cut-off frequency.
[13]Anritsu, model A3N1026, 8 kHz – 20 GHz bandwidth.
[14]Hewlett Packard, model HP8493C, 10 dB attenuation and 26.5 GHz cut-off frequency.
[15]NewFocus, model 1434-50, 25 GHz cut-off frequency.
[16]Hewlett Packard, model HP70000 system. The RF section HP70908A is employed.
[17]Picosecond Pulse Labs, model 5500, 80 kHz – 23 GHz bandwidth.

F Experimental Measurement Setups

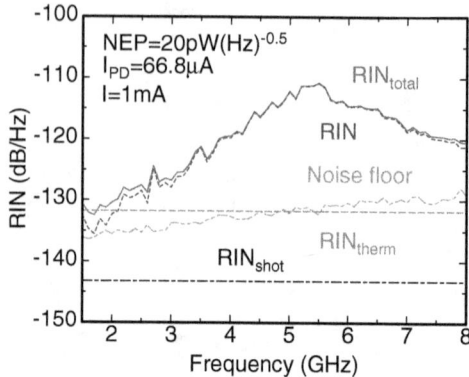

Figure F.6: RIN spectrum of a VCSEL at $I = 1\,\text{mA}$ along with the contributions of other noise sources in the measurement setup, such as the thermal and shot noise of the employed optical receiver.

measurement instruments are controlled by a PC using GBIP. Sample illumination and identification of individual devices are done equivalent to Fig. F.1.

F.4 RIN Measurements

Figure F.5 shows the experimental setup employed to measure the RIN spectra. It is similar to the setup used for the measurement of the small-signal modulation response curves. A wide-band LNA[18] is added after an optical receiver[19]. The optical receiver consists of an RF photodetector and a transimpedance amplifier with $19.5\,\text{dB}$ gain. The noise spectrum is recorded using a microwave spectrum analyzer[20]. The frequency responses of all RF cables, the bias-tee, the optical receiver, and the RF amplifier are measured and numerically subtracted from the measured noise spectrum. Figure F.6 depicts $\text{RIN}_{\text{total}}$ which contains not only the intensity noise of the laser but also other noise sources in the measurement system, such as the thermal noise and shot noise of the employed optical receiver. Therefore, the contribution of the noise floor of the system to $\text{RIN}_{\text{total}}$ is measured and numerically subtracted. The noise floor is measured while the VCSEL is not operated. It thus originates mainly from the thermal noise of the optical receiver $\langle |\Delta \tilde{I}_{\text{PD,therm}}(f)|^2 \rangle$ calculated by (3.54) and to an almost negligible extent from the shot

---
[18]Miteq, model AMF-3D-001100-25-13P, $30\,\text{dB}$ gain and $10\,\text{GHz}$ cut-off frequency.
[19]Picometrix, model AD-50xr, $12\,\text{GHz}$ cut-off frequency.
[20]Hewlett Packard, model HP70000 system. The RF section HP70908A is employed.

F.4 RIN Measurements

noise of the dark current. According to (3.57), the laser RIN is obtained by subtracting also the shot noise component RIN_{shot} calculated by (3.59). For the VCSEL of Fig. F.6, the received optical power generates an average photocurrent of $\langle I_{\text{PD}} \rangle = 66.8\,\mu\text{A}$.[21]

---

[21] The NEP of the optical receiver is $20\,\text{pW}/\sqrt{\text{Hz}}$ and the responsivity is $R_{\text{PD}} = 0.86\,\text{A/W}$, resulting in $\text{RIN}_{\text{therm}} = -131.8\,\text{dB/Hz}$ according to (3.58). According to (3.59), $\text{RIN}_{\text{shot}} = -143.2\,\text{dB/Hz}$. Both contributions are displayed in Fig. F.6.

F Experimental Measurement Setups

Appendix G

Cesium Absorption Spectra

This appendix explains the experimental setup and procedure applied to measure the D_1 absorption spectra of a Cs vapor cell which has been microfabricated by SAMLAB at EPFL[1]. The vapor cell contains, in addition to the Cs vapor, N_2 as buffer gas atmosphere. Absorption spectra of the Cs vapor cell measured at different cell temperatures are presented and discussed.

The experimental setup used to measure the absorption spectra of the Cs vapor cell is depicted in Fig. G.1. The VCSEL temperature is controlled using a temperature controller unit[2]. The vapor cell is mounted on a holder surrounded by heaters and a thermistor for temperature control. Behind the cell, a beam splitter is placed to divide the laser beam passing through the cell into two beams. For measuring optical spectra, one beam is coupled into a MMF connected to an optical spectrum analyzer[3]. The second beam is directed to a large-area silicon photodetector[4] to measure the transmitted optical power through the vapor cell using a power meter[5]. To avoid unwanted optical feedback, both sides of the MMF are angle-polished[6]. Moreover anti-reflection-coated lenses are employed and the photodetector is tilted by 5° with respect to the optical axis. The VCSEL is driven by a constant current generated by a laser driver[7]. All measurement instruments are controlled by a PC using GPIB.

PR-LIV characteristics of an inverted grating VCSEL at 45°C substrate temperature are shown in Fig. G.2 (left). The device remains polarization-stable from its threshold

---

[1] EPFL is a MAC-TFC project partner. See Sect. 2.3 and App. B to know more about the MAC-TFC consortium.
[2] ILX Lightwave, model LDT-5910B.
[3] ANDO Electric Corp., model AQ 6317, 0.05 nm spectral resolution.
[4] Newport Corp., model 818-ST.
[5] Newport Corp., model 1830-C.
[6] APC connector: the normal to the front surface of the optical fiber is tilted by 8 degrees with respect to the fiber axis.
[7] Keithley Instruments, model 2401, current resolution of 0.5 µA in the 10 mA range.

G Cesium Absorption Spectra

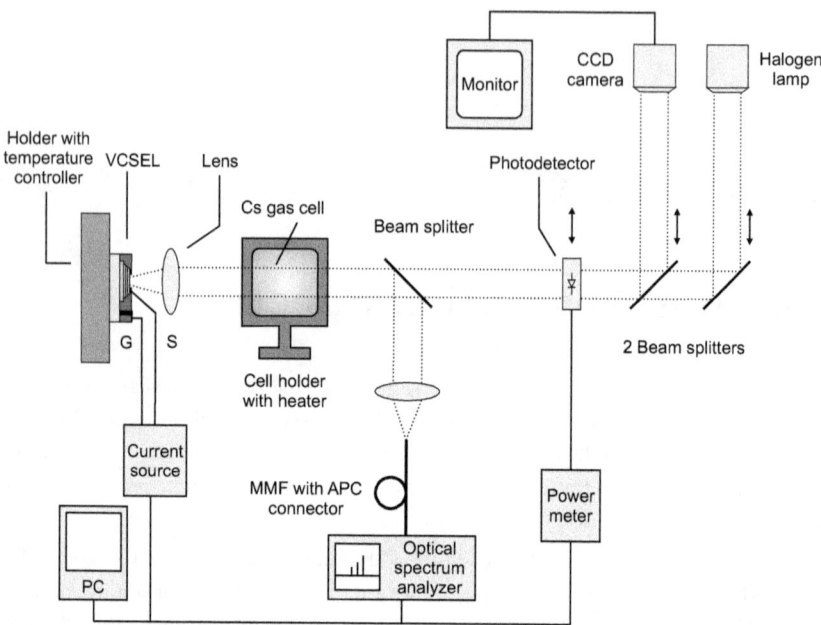

Figure G.1: Schematic drawing of the experimental setup used for measurements of the absorption spectra of the Cs gas cell.

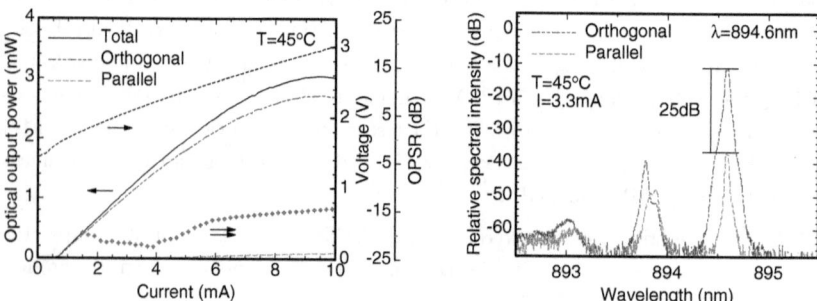

Figure G.2: Polarization-resolved operation characteristics of an inverted grating VCSEL with $D_a = 4.6\,\mu\text{m}$ at $T = 45°\text{C}$ (left) and its polarization-resolved spectra at $I = 3.3\,\text{mA}$ (right). The surface grating has 70 nm depth, 0.6 µm period, and 50% duty cycle. The laser has $x = 4\%$ and $N_p = 25$.

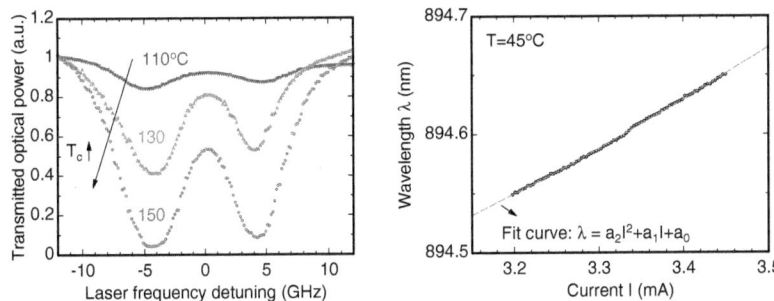

Figure G.3: Transmitted optical power as a function of frequency detuning for a Cs vapor cell at temperatures T_c of 110, 130 and 150°C (left). Emission wavelength λ as a function of current I along with a quadratic curve fit (dashed line) of the data points (right). $a_0 = 894.9$ nm, $a_1 = -578.9$ nm/mA and $a_2 = 148.12$ µm/(mA)^2 are fit parameters.

current of approximately 0.65 mA up to thermal roll-over with a maximum magnitude and an average of the OPSR of 22.6 dB and -18.5 dB, respectively. As expected from the simulation results shown in Sect. 4.4.3, the dominant polarization is orthogonal to the grating lines, resulting in a negative sign of the OPSR. Figure G.2 (right) depicts polarization-resolved spectra at 45°C. The target wavelength $\lambda_{D_1} = 894.6$ nm is reached at a current of 3.3 mA with an SMSR of about 28 dB and a peak-to-peak difference between the dominant and the suppressed polarization modes of almost 25 dB. This laser was utilized to measure the absorption spectra of the Cs vapor cell, as depicted in Fig G.3 (left). By sweeping the VCSEL current around $I = 3.3$ mA in steps of 0.5 µA, the laser emission wavelength is tuned by the red-shift effect over the hyperfine splitting frequency of 9.2 GHz around $\lambda = \lambda_{D_1}$. A wavelength tuning rate of approximately 400 pm/mA is achieved, as shown in Fig. G.3 (right). This corresponds to approximately 150 GHz/mA. From (3.16), (3.18), (3.25) and (3.27) the dissipated power can be approximated by

$$P_{\text{diss}} \approx R_s I^2 + (V_k - \text{SE})I + \text{SE}I_{\text{th}}. \tag{G.1}$$

In Sect. 3.4.2, the linear dependence of the emission wavelength on the dissipated power has been illustrated. Therefore the emission wavelength dependence on current can be approximated by a quadratic fit function as

$$\lambda = a_2 I^2 + a_1 I + a_0, \tag{G.2}$$

with a_2, a_1 and a_0 as fit parameters. The fit curve is added to Fig. G.3 (right) and shows a very good fit with the (I, λ) data points. The laser frequency detuning Δf_I in Fig. G.3

165

G Cesium Absorption Spectra

(left) is calculated from the λ versus I fit curve depicted in Fig. G.3 (right) by subtracting the D_1 emission frequency as

$$\Delta f = \frac{c}{\lambda} - \frac{c}{\lambda_{D_1}}. \tag{G.3}$$

Figure G.3 (left) depicts the transmitted optical power spectra through the Cs vapor cell at different cell temperatures T_c. Two absorption peaks start to appear at 110°C and they become more distinguishable at higher T_c. The pressure of the buffer gas in the alkali vapor cell was not available. However, according to the manufacturer, the pressure is expected to be somewhat high. The higher the pressure of the buffer gas in the cell, the broader are the atomic resonance lines [43]. Hence, in Fig. G.3 (left) only two absorption peaks appear rather than four peaks, as illustrated in Fig. 7.3 for a pure Cs vapor cell. Moreover, T_c (at which the absorption peaks become distinguishable) is higher than the MAC-TFC specification of 65 to 80°C. This also indicates that the pressure of the buffer gas is higher than required.

Appendix H

List of Acronyms

AFM	atomic force microscope
APC	angled physical contact
ASIC	application-specific integrated circuit
BNM	Bureau National de Métrologie
CB	conduction band
CBO	conduction band offset
CCD	charge-coupled device
CGPM	Conférence Générale des Poids et Mesures
CPT	coherent population trapping
CW	continuous wave
DBR	distributed Bragg reflector
DC	direct current
dpi	dots per inch
EC	European Commission
EEL	edge-emitting laser
EIT	electromagnetically induced transparency
FM	frequency modulation
FP7	seventh research framework programme of the EC
FWHM	full width at half maximum
GI	graded-index
GPIB	general purpose interface bus
HH	heavy hole
IM	intensity modulation
LH	light hole
LI	light–current

H List of Acronyms

LIV	light–current–voltage
LNA	low-noise amplifier
LNE	Laboratoire National de Métrologie et d'Essais
LP	linearly polarized
LPTF	Laboratoire Primaire du Temps et des Fréquences
LTCC	low-temperature co-fired ceramic
MAC	miniaturized (or MEMS) atomic clock
MAC-TFC	MEMS atomic clocks for timing, frequency control & communications
MBE	molecular beam epitaxy
MCEF	modulation current efficiency factor
MCXO	microcomputer-compensated crystal oscillator
MEMS	microelectromechanical systems
MMF	multi-mode fiber
MTTF	mean time to failure
NBS	National Bureau of Standards
NEP	noise-equivalent power
NIST	National Institute of Standards and Technology
NPL	National Physical Laboratory
NSD	noise spectral density
OCXO	oven-controlled crystal oscillator
OPSR	orthogonal polarization suppression ratio
PC	personal computer
PMMA	polymethyl methacrylate
PMR	professional mobile radio
ppm	pages per minute
PR-LI	polarization-resolved light–current
PR-LIV	polarization-resolved light–current–voltage
PTB	Physikalisch-Technische Bundesanstalt
PTP	precision time protocol
QB	quantum barrier
QW	quantum well
QWP	quarter-wave plate
RAFS	rubidium atomic frequency standards
RF	radio frequency
RIE	reactive-ion etching
RIN	relative intensity noise

SBNS	satellite-based navigation systems
SE	slope efficiency
SEM	scanning electron microscope
SG	signal–ground
SI	Système International d'Unités
SMSR	side-mode suppression ratio
SO	spin-orbit, split-off
SWaP	size, weight, and power
SYRTE	Systèmes de Référence Temps-Espace
TCXO	temperature-compensated crystal oscillator
TDLAS	tunable diode laser absorption spectroscopy
TEC	thermoelectric cooler
VBM	valence band maximum
VBO	valence band offset
VCSEL	vertical-cavity surface-emitting laser

H List of Acronyms

Appendix I

List of Symbols

I.1 Mathematical Operators, Special Functions and Constants

$\langle \bar{a} \rangle$ average, spatial average of a
d   differential operator
∂ partial differential operator
$J_{\bar{k}}$ Bessel function of order $\bar{k}$
c velocity of light; $c = 2.99792458 \cdot 10^8$ m/s [218]
e Euler number; $e = 2.718281828$
$\hbar$ $h/(2\pi)$; Planck's constant $h = 6.62606891(24) \cdot 10^{-34}$ J·s [218]
i   imaginary number; $\mathrm{i} = \sqrt{-1}$
k_B Boltzmann's constant; $k_\mathrm{B} = 1.3806504(24) \cdot 10^{-23}$ J/K
     $= 8.61734294 \cdot 10^{-5}$ eV/K [218]
q elementary charge; $q = 1.602176487(40) \cdot 10^{-19}$ C [218]
π pi; $\pi = 3.141592654$

I.2 Mathematical Symbols

a lattice parameter
$a_\mathrm{g+}, a_\mathrm{g-}$ curvatures of a gain spectrum at both sides of the gain peak
a_L lattice parameter of a grown layer
a_S lattice parameter of a substrate
$\bar{a}$ differential gain coefficient

I List of Symbols

$\bar{a}_c$	deformation potential constant
$\bar{a}_v$	Pikus–Bir deformation potential constant of hydrostatic strain
A	non-radiative recombination coefficient
A_a	active region area
A_{FN}	integral of $\tilde{S}_{FN}(f)$ over f
A_{ox}	area of the oxide aperture
$\bar{A}$	amplitude constant of the transfer function of the small-signal modulation response
$\bar{A}_0, \bar{A}_1$	amplitude constants of the RIN fit function
b	Pikus–Bir deformation potential constant of uniaxial strain
B	radiative recombination coefficient
$\bar{B}$	DC background noise
C	Auger recombination coefficient
C_a	combination of capacitances around the active region
C_{dep}	depletion layer capacitance
C_{diff}	diffusion capacitance
C_j	diode junction capacitance
C_{ox}	oxide layer capacitance
C_1	coefficient; wavelength shift with dissipated power at constant substrate temperature
C_2	coefficient; wavelength shift with varying substrate temperature at zero dissipated power
C_{11}, C_{12}	elastic stiffness coefficients
$\bar{C}$	contrast of the CPT signal
d	grating etch depth
d_a	thickness of the active layers
d_B	thickness of one period of a DBR
d_c	critical thickness of a grown layer
d_r	resist thickness
d_w	spatial width of one-dimensional symmetric quantum well
d'_r	resist thickness in units of nm
D	D-factor
D_a	active diameter of a VCSEL
D_{grat}	diameter of circular area over which a grating pattern extends
D_m	mesa diameter at the oxide layer
D_{ring}	diameter of contact ring opening

I.2 Mathematical Symbols

E	energy eigenstate in a quantum well
E_a	failure activation energy
E_e	electron energy
E_g	bandgap energy
$E_{g,b}$	bandgap energy of a quantum barrier
$E_{g,bow}$	bandgap bowing parameter
$E_{g,renorm}$	renormalized bandgap energy
$E_{g,strained}$	strained bandgap energy
$E_{g,unstrained}$	unstrained bandgap energy
$E_{g,w}$	bandgap energy of a quantum well
E_g^X	bandgap energy at the X-point
E_g^Γ	bandgap energy at the Γ-point
$\bar{E}$	electric field
$\bar{E}_0$	electric field amplitude constant
$\bar{E}_m$	electric field of light emission of an intensity-modulated VCSEL
$\tilde{E}$	Fourier transform of the electric field of laser emission
δE_C	energy shift of direct conduction band minimum
δE_g	bandgap correction factor
δE_{HH}	band-edge energy of heavy-hole sub-band
δE_{LH}	band-edge energy of light-hole sub-band
ΔE_C	relative band offset of the conduction band
ΔE_V	relative band offset of the valence band
f	modulation frequency
f_e	exposure distribution function of the electron beam
$f_{max,d}$	theoretical damping-limited maximum 3 dB corner frequency
$f_{max,t}$	thermally-limited maximum 3 dB corner frequency
f_{meas}	measured frequency of a clock or a frequency standard
f_{nom}	nominal frequency of a clock or a frequency standard
f_p	3 dB corner frequency of the parasitic transfer function
f_{peak}	frequency at the peak of the modulation response
f_r	resonance frequency
$f_{r,max}$	maximum resonance frequency
f_0	optical carrier frequency
f_{3dB}	3 dB corner frequency of the modulation response
$f_{3dB,max}$	maximum 3 dB corner frequency
Δf	noise bandwidth

I List of Symbols

Δf_{I}	laser detuning frequency
Δf_{L}	full linewidth at half maximum of the laser emission
$\Delta f_{\mathrm{L_0}}$	residual linewidth
F	quantum number associated with $\mathbf{F}$
$\mathbf{F}$	total atomic angular momentum; $\mid \mathbf{F} \mid^2 = F(F+1)\hbar^2$
F_z	z component of $\mathbf{F}$
ΔF	change in the quantum number F
g	optical gain coefficient
g_{p}	peak gain coefficient
g_{para}	material threshold gain of the fundamental mode polarized parallel to the grating lines
g_{orth}	material threshold gain of the fundamental mode polarized orthogonal to the grating lines
g_{th}	threshold gain
$\bar{g}$	gain constant
I	bias current
I	quantum number associated with $\mathbf{I}$ (only in App. A)
$\mathbf{I}$	total nuclear spin angular momentum; $\mid \mathbf{I} \mid^2 = I(I+1)\hbar^2$
I_{PD}	detector photocurrent
I_{th}	threshold current
I_0	spectral power intensity of the optical carrier
I_{+1}	spectral power intensity of the upper first-order modulation sideband
I_{-1}	spectral power intensity of the lower first-order modulation sideband
$\Delta \tilde{I}_{\mathrm{PD}}$	Fourier transform of photocurrent noise
$\Delta \tilde{I}_{\mathrm{PD,laser}}$	Fourier transform of the photocurrent noise caused by laser noise
$\Delta \tilde{I}_{\mathrm{PD,shot}}$	Fourier transform of the photocurrent noise caused by shot noise
$\Delta \tilde{I}_{\mathrm{PD,therm}}$	Fourier transform of the photocurrent noise caused by thermal noise
J	current density
J	quantum number associated with $\mathbf{J}$ (only in App. A)
$\mathbf{J}$	total electron angular momentum; $\mid \mathbf{J} \mid^2 = J(J+1)\hbar^2$
J_0	current density of a reference VCSEL
J_{th}	threshold current density
J_z	z component of $\mathbf{J}$
ΔJ	change in the quantum number J
k	wavenumber
k_{n}	normalization factor of f_{e}

I.2 Mathematical Symbols

K	K-factor
l	azimuthal order of LP_{lp} mode
l_{eff}	effective penetration depth into a DBR
$l_{\text{eff,b}}$	effective penetration depth into the bottom DBR
$l_{\text{eff,t}}$	effective penetration depth into the top DBR
L	metal track inductance
L	quantum number associated with $\mathbf{L}$ (only in App. A)
$\mathbf{L}$	total orbital angular momentum; $\mid \mathbf{L} \mid^2 = L(L+1)\hbar^2$
L_c	laser inner cavity length
$L_{c,\text{eff}}$	effective cavity length
L_z	z component of $\mathbf{L}$
ΔL	change in the quantum number L
m	laser mode order
m_d	grating diffraction order
m_F	quantum number associated with $\mathbf{F}$; $F_z = m_F \hbar$
m_J	quantum number associated with $\mathbf{J}$; $J_z = m_J \hbar$
m_L	quantum number associated with $\mathbf{L}$; $L_z = m_L \hbar$
m_S	quantum number associated with $\mathbf{S}$; $S_z = m_S \hbar$
m_b^*	effective mass of carriers in a quantum barrier
m_e^*	effective mass of electrons
m_{HH}^*	effective mass of heavy holes
m_{LH}^*	effective mass of light holes
m_w^*	effective mass of carriers in a quantum well
Δm_F	change in the quantum number m_F
M	frequency modulation index
M_a	number of active layers
M_p	parasitic transfer function of the equivalent-circuit model
M_t	transfer function of the small-signal modulation response
$\bar{M}$	number of relative frequency offsets used to calculate the Allan deviation
n	carrier density
n_Q	principal quantum number
n_t	transparency carrier density
n_{th}	threshold carrier density
n_0	carrier density at the operating point
$\bar{n}$	refractive index
$\bar{n}_{\text{gr}}$	group refractive index

I List of Symbols

$\bar{n}_t$	linearization transparency carrier density
$\Delta \bar{n}_B$	refractive index step in a binary DBR
N_p	number of top Bragg mirror pairs
$\bar{N}_{MAC}$	overall noise of the MAC at the RF modulation signal
$\bar{N}_{mod}$	number of transverse modes of a multi-mode VCSEL
$\Delta \tilde{N}$	Fourier transform of photon density noise
p	radial order of LP$_{lp}$ mode
P	optical output power
P_{diss}	dissipated electrical power
P_{orth}	power in the polarization orthogonal to the grating lines
P_{par}	power in the polarization parallel to the grating lines
δP	optical output power noise
$\Delta \tilde{P}$	Fourier transform of δP
Q	quality factor
r	radial position from center of the primary electron beam
R	intensity modulation index
R_a	oxide aperture resistance
R_b	intensity reflection coefficient of a lossless bottom DBR
$R_{b\alpha}$	intensity reflection coefficient of a lossy bottom DBR
R_m	mirror resistance
R_{PD}	photodetector responsivity
R_s	differential series resistance
R_t	intensity reflection coefficient of a lossless top DBR
$R_{t\alpha}$	intensity reflection coefficient of a lossy top DBR
R_{th}	thermal resistance
S	sideband asymmetry factor
S	quantum number associated with **S** (only in App. A)
S	total spin angular momentum; $\mid \mathbf{S} \mid^2 = S(S+1)\hbar^2$
S_z	z component of **S**
S_{11}	reflection scattering parameter at the input of the equivalent-circuit model
$\tilde{S}$	amplitude of the CPT signal
$\tilde{S}_{FN}$	spectral power density of laser frequency noise
ΔS	change in the quantum number S
t	time
t_{cap}	cap layer thickness
t_{etch}	grating etch time

I.2 Mathematical Symbols

t_i	time measurement
Δt	phase deviation
T	ambient (or substrate) temperature
T_{int}	average temperature of the inner cavity of a VCSEL
$T_{\text{int},0}$	average temperature of the inner cavity of a reference VCSEL
T_{inv}	inversion temperature
ΔT	temperature increase inside a VCSEL
v_{gr}	group velocity
V	voltage across a VCSEL
V_{a}	active region volume
V_{acc}	acceleration voltage of electron beam
V_{k}	kink voltage
V_{p}	photon volume
$V_{R_{\text{a}}}$	small-signal modulating voltage reaching the active region of a VCSEL
V_{s}	small-signal modulating voltage generated by the RF source
V_{w}	potential depth of a quantum well
V'_{acc}	acceleration voltage of electron beam in units of kV
VBO_{b}	valence band offset of a quantum barrier
VBO_{bow}	bowing parameter a valence band offset
VBO_{w}	valence band offset of a quantum well
W_{air}	grating groove width
W_{sem}	grating ridge width
x_{aging}	aging acceleration factor
$\bar{y}, \bar{y}_i$	relative frequency offset
z	atomic quantization axis
$z_{i\text{h}}$	longitudinal position of the end of an active layer
$z_{i\text{l}}$	longitudinal position of the beginning of an active layer
$\bar{z}$	depth of electron beam in resist
Z	input impedance of the electrical equivalent-circuit model
Z_0	characteristic impedance of the measurement system; usually $Z_0 = 50\,\Omega$

I.3 Greek Symbols

α	intensity attenuation coefficient
α_a	intrinsic losses in the active layers
α_f	electron-beam broadening factor
α_H	Henry factor or linewidth enhancement factor
α_i	intrinsic losses in the passive layers
α_m	mirror losses
$\hat{\alpha}$	fit parameter of bandgap energy versus absolute temperature
β_b	electron-beam broadening caused by backward-scattered electrons
β_f	electron-beam broadening caused by forward-scattered electrons
β_{sp}	spontaneous emission factor
β_0	maximum achievable resolution of electron-beam lithography
$\hat{\beta}$	fit parameter of bandgap energy versus absolute temperature
γ	damping coefficient
$\bar{\gamma}$	complex propagation constant
γ_0	damping coefficient offset
Γ	confinement factor
Γ_r	relative confinement factor or gain enhancement factor
Γ_t	transverse confinement factor
Γ_z	longitudinal confinement factor
Δ_{SO}	spin-orbit splitting energy
$\bar{\varepsilon}$	gain compression parameter
$\varepsilon_\parallel$	biaxial strain in the substrate plane
$\varepsilon_\perp$	uniaxial strain perpendicular to the substrate plane
η_c	conversion (or wallplug) efficiency
η_d	differential quantum efficiency
η_E	electron back-scattering coefficient
η_I	current injection efficiency
η_{PD}	photodetector quantum efficiency
$\hat{\eta}_c$	maximum conversion (or wallplug) efficiency
$\tilde{\eta}_{d,t}$	photonic quantum efficiency
θ_{SL}	far-field angle of diffraction side-lobes
λ	emission wavelength
λ_B	Bragg wavelength
λ_c	thermal conductivity
$\lambda_{c,0}$	thermal conductivity at room temperature

I.3 Greek Symbols

λ_{D_1}	wavelength of the Cs D_1 absorption line
λ_g	bandgap wavelength
λ_{mat}	material wavelength
λ_p	gain peak wavelength
Λ	grating period
$\delta\lambda_g$	offset between emission wavelength and gain peak wavelength
σ_P	Poisson's ratio
$\sigma_{\bar{y}}$	Allan deviation
τ	measurement period for Allan deviation
τ_p	photon lifetime
τ_{sp}	spontaneous recombination lifetime
$\tau_{sp,n}$	non-radiative spontaneous recombination lifetime
$\tau_{sp,r}$	radiative spontaneous recombination lifetime
ϕ	relative phase between frequency and intensity modulations
ϕ_b	phase shift of the electric field at the interface of inner cavity and bottom DBR
ϕ_t	phase shift of the electric field at the interface of inner cavity and top DBR
ψ	electron wave function
ω	angular frequency

I List of Symbols

Publications

Conference Proceedings

1. R. Michalzik, J.M. Ostermann, **A. Al-Samaneh**, D. Wahl, F. Rinaldi, and P. Debernardi, "Polarization-stable VCSELs for optical sensing and communications" (invited), in Proc. (CD ROM) *The 14th OptoElectronics and Communications Conf., OECC 2009*, paper TuC2, two pages. Hong Kong, China, July 2009, DOI: 10.1109/OECC.2009.5219064.

2. **A. Al-Samaneh**, S. Renz, A. Strodl, W. Schwarz, D. Wahl, and R. Michalzik, "Polarization-stable single-mode VCSELs for Cs-based MEMS atomic clock applications", in *Semiconductor Lasers and Laser Dynamics IV*, K.P. Panayotov, M. Sciamanna, A.A. Valle, R. Michalzik (Eds.), Proc. SPIE 7720, pp. 772006-1–14, 2010, DOI: 10.1117/12.853181.

3. L. Bimboes, F. Gruet, C. Affolderbach, R. Matthey, G. Mileti, **A. Al-Samaneh**, D. Wahl, and R. Michalzik, "Spectral characterization of 894 nm VCSELs for MEMS atomic clocks", Atelier à Arc-et-Senans, Les microtechniques dans le quotidien, one-page abstract and poster. Saline Royale d'Arc-et-Senans, France, Sept. 2010.

4. **A. Al-Samaneh**, S. Renz, D. Wahl, and R. Michalzik, "Small-signal analysis and polarization stability performance of single-mode VCSELs for MEMS atomic clock applications", in Proc. *Sixth Joint Symposium on Opto- and Microelectronic Devices and Circuits, SODC 2010*, pp. 61–64. Berlin, Germany, Oct. 2010.

5. **A. Al-Samaneh**, M.T. Haidar, D. Wahl, and R. Michalzik, "Polarization-stable single-mode VCSELs for Cs-based miniature atomic clocks", in Online Digest *Conf. on Lasers and Electro-Optics Europe, CLEO/Europe 2011*, paper CB.P.23, one page. Munich, Germany, May 2011, DOI: 10.1109/CLEOE.2011.5942762.

6. F. Gruet, L. Bimboes, D. Miletic, C. Affolderbach, G. Mileti, **A. Al-Samaneh**, D. Wahl, and R. Michalzik, "Spectral characterisation of VCSELs emitting at 894.6 nm for CPT-based miniature atomic clocks", in Online Digest *Conf. on Lasers and Electro-Optics Europe, CLEO/Europe 2011*, paper CB.P.27, one page. Munich, Germany, May 2011, DOI: 10.1109/CLEOE.2011.5942636.

7. R. Boudot, X. Liu, E. Kroemer, P. Abbé, N. Passilly, S. Galliou, R.K. Chutani, V. Giordano, C. Gorecki, **A. Al-Samaneh**, D. Wahl, and R. Michalzik, "Characterization of compact CPT clocks based on a Cs-Ne microcell", in Proc. *2012 European Frequency and Time Forum, EFTF*, pp. 79–82. Gothenburg, Sweden, Apr. 2012, DOI: 10.1109/EFTF.2012.6502338.

8. A. Kern, **A. Al-Samaneh**, D. Wahl, and R. Michalzik, "10 Gbit/s bidirectional multimode data link using monolithically integrated VCSEL–PIN transceiver devices", in Proc. (USB flash drive) *38th Europ. Conf. on Opt. Commun., ECOC 2012*, paper We.1.E.2, three pages. Amsterdam, The Netherlands, Sept. 2012, DOI: 10.1364/ECEOC.2012.We.1.E.2.

9. M.J. Miah, **A. Al-Samaneh**, D. Wahl, and R. Michalzik, "Dynamic characteristics of inverted grating relief VCSELs for Cs-based microscale atomic clocks", in Online Digest *Conf. on Lasers and Electro-Optics Europe, CLEO/Europe 2013*, paper CB-8.4-THU, one page. Munich, Germany, May 2013.

Journal Papers

1. **A. Al-Samaneh**, M. Bou Sanayeh, S. Renz, D. Wahl, and R. Michalzik, "Polarization control and dynamic properties of VCSELs for MEMS atomic clock applications", *IEEE Photon. Technol. Lett.*, vol. 23, pp. 1049–1051, 2011, DOI: 10.1109/LPT.2011.2151853.

2. **A. Al-Samaneh**, M. Bou Sanayeh, W. Schwarz, D. Wahl, and R. Michalzik, "Vertical-cavity lasers for miniaturized atomic clocks", *SPIE Newsroom*, Aug. 18, 2011, DOI: 10.1117/2.1201108.003824.

3. A. Kern, S. Paul, D. Wahl, **A. Al-Samaneh**, and R. Michalzik, "Single-fiber bidirectional optical data links with monolithic transceiver chips" (invited), *Advances in Optical Technologies*, Special Issue on *Recent Advances in Semiconductor Surface-Emitting Lasers*, Article ID 729731, 8 pages, 2012, DOI: 10.1155/2012/729731.

4. **A. Al-Samaneh**, M. Bou Sanayeh, M.J. Miah, W. Schwarz, D. Wahl, A. Kern, and R. Michalzik, "Polarization-stable vertical-cavity surface-emitting lasers with

inverted grating relief for use in microscale atomic clocks", *Appl. Phys. Lett.*, vol. 101, pp. 171104-1–4, 2012, DOI: 10.1063/1.4764010.

5. F. Gruet, **A. Al-Samaneh**, E. Kroemer, L. Bimboes, D. Miletic, C. Affolderbach, D. Wahl, R. Boudot, G. Mileti, and R. Michalzik, "Metrological characterization of custom-designed 894.6 nm VCSELs for miniature atomic clocks", *Opt. Express*, vol. 21, pp. 5781–5792, 2013, DOI: 10.1364/OE.21.005781.

6. M.J. Miah, **A. Al-Samaneh**, A. Kern, D. Wahl, P. Debernardi, and R. Michalzik, "Fabrication and characterization of low-threshold polarization-stable VCSELs for Cs-based miniaturized atomic clocks", *IEEE J. Select. Topics Quantum Electron.*, vol. 19, 1701410, 10 pages, 2013, DOI: 10.1109/JSTQE.2013.2247697.

7. A. Kern, **A. Al-Samaneh**, D. Wahl, and R. Michalzik, "Monolithic VCSEL–PIN photodiode integration for bidirectional optical data transmission", *IEEE J. Select. Topics Quantum Electron.*, vol. 19, 6100313, 13 pages, 2013, DOI: 10.1109/JSTQE.2013.2245102.

Publications

Bibliography

[1] T. Quinn (Ed.), "Special issue: Fifty years of atomic timekeeping: 1955 to 2005", *Metrologia*, vol. 42, pp. S1–S153, 2005.

[2] F. Riehle, *Frequency Standards: Basics and Applications*. Weinheim, Germany: WILEY-VCH Verlag, 2004.

[3] S. Knappe, V. Shah, P. Schwindt, L. Hollberg, J. Kitching, L.A. Liew, and J. Moreland, "A microfabricated atomic clock", *Appl. Phys. Lett.*, vol. 85, pp. 1460–1462, 2004.

[4] R. Lutwak, J. Deng, W. Riley, M. Varghese, J. Leblanc, G. Tepolt, M. Mescher, D.K. Serkland, K.M. Geib, and G.M. Peake, "The chip-scale atomic clock – low-power physics package", in Proc. *36th Annual Precise Time and Time Interval (PTTI) Meeting*, pp. 339–354. Washington DC, USA, Dec. 2004.

[5] C. Gorecki, M. Hasegawa, N. Passilly, R.K. Chutani, P. Dziuban, S. Gailliou, and V. Giordano, "Towards the realization of the first European MEMS atomic clock", in Proc. *2009 IEEE/LEOS International Conference on Optical MEMS and Nanophotonics*, pp. 47–48. Clearwater, USA, Aug. 2009.

[6] http://www.mac-tfc.eu, last visited October 2013.

[7] H.R. Gray, R.M. Whitley, and C.R. Stroud, Jr., "Coherent trapping of atomic populations", *Opt. Lett.*, vol. 3, pp. 218–220, 1978.

[8] N. Cyr, M. Têtu, and M. Breton, "All-optical microwave frequency standard: a proposal", *IEEE Transactions on Instrumentation and Measurement*, vol. 42, pp. 640–649, 1993.

[9] D.K. Serkland, G.M. Peake, K.M. Geib, R. Lutwak, R.M. Garvey, M. Varghese, and M. Mescher, "VCSELs for atomic clocks", in *Vertical-Cavity Surface-Emitting Lasers X*, C. Lei and K.D. Choquette (Eds.), Proc. SPIE 6132, pp. 613208-1–11, 2006.

[10] D.K. Serkland, K.M. Geib, G.M. Peake, R. Lutwak, A. Rashed, M. Varghese, G. Tepolt, and M. Prouty, "VCSELs for atomic sensors", in *Vertical-Cavity Surface-Emitting Lasers XI*, K.D. Choquette and J.K. Guenter (Eds.), Proc. SPIE 6484, pp. 648406-1–10, 2007.

[11] G. Verschaffelt, K. Panajotov, J. Albert, B. Nagler, M. Peeters, J. Danckaert, I. Veretennicoff, and H. Thienpont, "Polarisation switching in vertical-cavity surface-emitting lasers: from experimental observations to applications", *Opto-Electron. Rev.*, vol. 9, pp. 257–268, 2001.

[12] J.M. Ostermann and R. Michalzik, "Polarization Control of VCSELs", Chap. 5 in *VCSELs — Fundamentals, Technology and Applications of Vertical-Cavity Surface-Emitting Lasers*, R. Michalzik (Ed.), Springer Series in Optical Sciences, vol. 166, pp. 147–179. Berlin, Germany: Springer-Verlag, 2013.

[13] G.P. Bava, P. Debernardi, and L. Fratta, "Three-dimensional model for vectorial fields in vertical-cavity surface-emitting lasers", *Phys. Rev. A*, vol. 63, pp. 023816-1–13, 2001.

[14] J. Jespersen and J. Fitz-Randolph, *From Sundials To Atomic Clokcs – Understanding Time and Frequency*. Washington DC, USA: Natl. Inst. Stand. Technol., Monograph 155, 1999.

[15] M.A. Lombardi, "Fundamentals of Time and Frequency", Chap. 17 in *The Mechatronics Handbook*, R.H. Bishop (Ed.), 18 pages. New York, USA: CRC Press LLC, 2002.

[16] http://www.nist.gov/pml/general/time/early.cfm, last visited October 2013.

[17] J.R. Vig, "Quartz crystal resonators and oscillators for frequency control and timing applications – a tutorial", edition: Rev. 8.5.4.4, Apr. 2012. Web page: http://www.ieee-uffc.org/frequency_control/teaching_Vig.asp, last visited October 2013.

[18] R. Filler and J. Vig, "Resonators for the microcomputer-compensated crystal oscillator", in *43rd Annual Symposium on Frequency Control*, pp. 8–15. Denver, USA, May–Jun. 1989.

[19] L. Essen and J.V.L. Parry, "The caesium resonator as a standard of frequency and time", *Phil. Trans. Roy. Soc.*, vol. 250, pp. 45–69, 1957.

[20] P. Forman, "Atomichron®: The atomic clock from concept to commercial product", *Proc. IEEE*, vol. 73, no. 7, pp. 1181–1204, 1985.

[21] W.A. Mainberger and A. Orenberg, "The Atomichron® – atomic frequency standard: Operation and performance", in *IRF Nat. Conv. Rec.*, pt. 1, pp. 14–18, 1958.

[22] A. Bauch, K. Dorenwendt, B. Fischer, T. Heindorff, E.K. Müller, and R. Schröder, "CS2: The PTB's new primary clock", *IEEE Trans. Instrum. Meas.*, vol. IM-36, pp. 613–616, 1987.

[23] A. Clairon, C. Salomon, S. Guellati, and W.D. Phillips, "Ramsey resonance in a Zacharias fountain", *Europhys. Lett.*, vol. 16, pp. 165–170, 1991.

[24] C. Vian, P. Rosenbusch, H. Marion, S. Bize, L. Cacciapuoti, S. Zhang, M. Abgrall, D. Chambon, I. Maksimovic, P. Laurent, G. Santarelli, A. Clairon, A. Luiten, M. Tobar, and C. Salomon, "BNM-SYRTE fountains: Recent results", *IEEE Trans. Instrum. Meas.*, vol. 54, pp. 833–836, 2005.

Bibliography

[25] S. Weyers, B. Lipphardt, and H. Schnatz, "Reaching the quantum limit in a fountain clock using a microwave oscillator phase locked to an ultrastable laser", *Phys. Rev. A*, vol. 79, pp. 031803-1–4, 2009.

[26] S.R. Jefferts, T.P. Heavner, T.E. Parker, and J.H. Shirley, "NIST cesium fountains: current status and future prospects", in *Time and Frequency Metrology*, R.J. Jones (Ed.), Proc. SPIE 6673, pp. 667309-1–9, 2007.

[27] T.P. Heavner, T.E. Parker, J.H. Shirley, and S.R. Jefferts, "NIST F1 and F2", in Proc. *7th Symposium of Frequency Standards and Metrology*, pp. 299–307. Pacific Grove, USA, Oct. 2008.

[28] T.P. Heavner, T.E. Parker, J.H. Shirley, P. Kunz, and S.R. Jefferts, "NIST F1 and F2", in Proc. *42nd Annual Precise Time and Time Interval (PTTI) Meeting*, pp. 457–463. Reston, USA, Nov. 2010.

[29] S. Knappe, "MEMS Atomic Clocks", Chap. 3.18 in *Comprehensive Microsystems*, Y. Gianchandani, O. Tabata, and H. Zappe (Eds.), pp. 571–612. Amsterdam, The Netherlands: Elsevier, 2008.

[30] T. McClelland, I. Pascaru, I. Shtaerman, C. Stone, C. Szekely, J. Zacharski, N.D. Bhaskar, "Subminiature rubidium frequency standard: Manufacturability and performance results from production units", in Proc. *IEEE International Frequency Control Symposium*, pp. 39–52. San Francisco, USA, May 1995.

[31] P. Rochat, B. Leuenberger, and X. Stehlin, "A new synchronized ultra miniature rubidium oscillator", in Proc. *2002 IEEE International Frequency Control Symposium and PDA Exhibition*, pp. 451–454. New Orleans, USA, Dec. 2002.

[32] W. D. Jones, IEEE Spectrum, March 2011, http://spectrum.ieee.org/semiconductors/devices/chipscale-atomic-clock, last visited October 2013.

[33] http://www.symmetricom.com/CSAC/, last visited October 2013.

[34] M. Stähler, R. Wynands, S. Knappe, J. Kitching, L. Hollberg, A. Taichenachev, and V. Yudin, "Coherent population trapping resonances in thermal ^{85}Rb vapor: D_1 versus D_2 line excitation", *Opt. Lett.*, vol. 27, pp. 1472–1474, 2002.

[35] R. Lutwak, D. Emmons, T. English, W. Riley, A. Duwel, M. Varghese, D.K. Serkland, and G.M. Peake, "The chip-scale atomic clock – recent development progress", in Proc. *35th Annual Precise Time and Time Interval (PTTI) Meeting*, pp. 467–478. San Diego, USA, Dec. 2003.

[36] E. Arimondo and G. Orriols, "Nonabsorbing atomic coherences by coherent two-photon transitions in a three-level optical pumping", *Lettere al nuovo cimento*, vol. 17, pp. 333–338, 1976.

[37] S.E. Harris, J.E. Field, and A. Imamoğlu, "Nonlinear optical processes using electromagnetically induced transparency", *Phys. Rev. Lett.*, vol. 67, pp. 1107–1110, 1990.

[38] L. Nieradko, C. Gorecki, A. Douahi, V. Giordano, J.C. Beugnot, J. Dziuban, and M. Moraja, "New approach of fabrication and dispensing of micromachined cesium vapor cell", *J. Micro/Nanolithography, MEMS, and MOEMS*, vol. 7, pp. 033013-1–6, 2008.

[39] M. Hasegawa, R.K. Chutani, C. Gorecki, R. Boudot, P. Dziuban, V. Giordano, S. Clatot, and L. Mauri, "Microfabrication of cesium vapor cells with buffer gas for MEMS atomic clocks", *Sensors Actuat. A: Phys.*, vol. 167, pp. 594–601, 2011.

[40] R.H. Dicke, "The effect of collisions upon the Doppler width of spectral lines", *Phys. Rev.*, vol. 89, pp. 472–473, 1953.

[41] E.C. Beaty and P.L. Bender, "Narrow hyperfine absorption lines of Cs^{133} in various buffer gases", *Phys. Rev.*, vol. 112, pp. 450–451, 1958.

[42] J.C. Camparo and J.G. Coffer, "Conversion of laser phase noise to amplitude noise in a resonant atomic vapor: The role of laser linewidth", *Phys. Rev. A*, vol. 59, pp. 728–735, 1999.

[43] G.A. Pitz, D.E. Wertepny, and G.P. Perram, "Pressure broadening and shift of the cesium D_1 transition by the noble gases and N_2, H_2, HD, D_2, CH_4, C_2H_6, CF_4, and ^3He", *Phys. Rev. A*, vol. 80, pp. 062718-1–8, 2009.

[44] M. Huang and J. Camparo, "The influence of laser polarization variations on CPT atomic clock signals", in Proc. *2011 Joint Conference of the IEEE International Frequency Control and the European Frequency and Time Forum (FCS)*, pp. 951–954. San Francisco, USA, May 2011.

[45] K. Panajotov and F. Prati, "Polarization Dynamics of VCSELs", Chap. 6 in *VCSELs — Fundamentals, Technology and Applications of Vertical-Cavity Surface-Emitting Lasers*, R. Michalzik (Ed.), Springer Series in Optical Sciences, vol. 166, pp. 181–231. Berlin, Germany: Springer-Verlag, 2013.

[46] B.S. Mathur, H. Tang, and W. Happer, "Light shifts in the alkali atoms", *Phys. Rev.*, vol. 171, pp. 11–19, 1968.

[47] M. Zhu and J. DeNatale, "Application of reduced light shift optical pumping method to chip scale atomic clock", in Proc. *2009 Joint Meeting of the European Frequency and Time Forum (EFTF) and the IEEE International Frequency Control Symposium (FCS)*, pp. 1183–1186. Besançon, France, Apr. 2009.

[48] R. Boudot, P. Dziuban, M. Hasegawa, R.K. Chutani, S. Galliou, V. Giordano, and C. Gorecki, "Coherent population trapping resonances in CsNe vapor microcells for miniature clocks applications", *J. Appl. Phys.*, vol. 109, pp. 014912-1–11, 2011.

[49] V. Shah, V. Gerginov, P.D.D. Schwindt, S. Knappe, L. Hollberg, and J. Kitching, "Continuous light-shift correction in modulated coherent population trapping clocks", *Appl. Phys. Lett.*, vol. 89, pp. 151124-1–3, 2006.

Bibliography

[50] J. Vanier, R. Kunski, N. Cyr, J.Y. Savard, and M. Têtu, "On hyperfine frequency shifts caused by buffer gases: Application to the optically pumped passive rubidium frequency standard", *J. Appl. Phys.*, vol. 53, pp. 5387–5391, 1982.

[51] D. Miletic, P. Dziuban, R. Boudot, M. Hasegawa, R.K. Chutani, G. Mileti, V. Giordano, and C. Gorecki, "Quadratic dependence on temperature of Cs 0–0 hyperfine resonance frequency in single Ne buffer gas microfabricated vapour cell", *Electron. Lett.*, vol. 46, pp. 1069–1071, 2010.

[52] D. Serkland and R. Lutwak, "Atomic clocks throw down the gauntlet to VCSEL makers", *Compound Semiconductor*, vol. 13, pp. 14–16, 2007.

[53] J.R. Vig, "Military applications of high accuracy frequency standards and clocks", *IEEE Trans. Ultrason. Ferroelectr. Freq. Control.*, vol. 40, pp. 522–527, 1993.

[54] http://www.nist.gov/el/isd/ieee/tutorials-1588.cfm, last visited October 2013.

[55] http://www.oscilloquartz.com, last visited October 2013.

[56] H-P.A. Ketterling, *Introduction to Digital Professional Mobile Radio*. Boston, USA: Artech House, 2004.

[57] R. Michalzik (Ed.), *VCSELs — Fundamentals, Technology and Applications of Vertical-Cavity Surface-Emitting Lasers*, Springer Series in Optical Sciences, vol. 166. Berlin, Germany: Springer-Verlag, 2013.

[58] H. Li and K. Iga (Eds.), *Vertical-Cavity Surface-Emitting Laser Devices*. Berlin, Germany: Springer-Verlag, 2003.

[59] C. Wilmsen, H. Temkin, and L.A. Coldren (Eds.), *Vertical-Cavity Surface-Emitting Lasers: Design, Fabrication, Characterization, and Applications*. Cambridge: Cambridge University Press, 1999.

[60] K. Iga, F. Koyama, and S. Kinoshita, "Surface emitting semiconductor lasers", *IEEE J. Quantum Electron.*, vol. 24, pp. 1845–1855, 1988.

[61] H. Soda, K. Iga, C. Kitahara, and Y. Suematsu, "GaInAsP/InP surface emitting injection lasers", *Jpn. J. Appl. Phys.*, vol. 18, pp. 2329–2330, 1979.

[62] U. Fiedler and K.J. Ebeling, "Design of VCSEL's for feedback insensitive data transmission and external cavity active mode-locking", *IEEE J. Select. Topics Quantum Electron.*, vol. 1, pp. 442–450, 1995.

[63] H. Roscher, F. Rinaldi, and R. Michalzik, "Small-pitch flip-chip-bonded VCSEL arrays enabling transmitter redundancy and monitoring in 2-D 10-Gbit/s space-parallel fiber transmission", *IEEE J. Select. Topics Quantum Electron.*, vol. 13, pp. 1279–1289, 2007.

[64] R.L. Thornton, R.D. Burnham, and W. Streifer, "High reflectivity GaAs-AlGaAs mirrors fabricated by metalorganic chemical vapor deposition", *Appl. Phys. Lett.*, vol. 45, pp. 1028–1030, 1984.

[65] P.L. Gourley and T.J. Drummond, "Single crystal, epitaxial multilayers of AlAs, GaAs, and $Al_xGa_{1-x}As$ for use as optical interferometric elements", *Appl. Phys. Lett.*, vol. 49, pp. 489–491, 1986.

[66] R. Michalzik, "VCSEL Fundamentals", Chap. 2 in *VCSELs — Fundamentals, Technology and Applications of Vertical-Cavity Surface-Emitting Lasers*, R. Michalzik (Ed.), Springer Series in Optical Sciences, vol. 166, pp. 19–75. Berlin, Germany: Springer-Verlag, 2013.

[67] R. Michalzik, W. Schmid, S.W.Z. Mahmoud, and R. Jäger, *REFLEX Transfer Matrix Computer Program*. Ulm University, Institute of Optoelectronics.

[68] D.L. Huffaker, D.G. Deppe, K. Kumar, and T.J. Rogers, "Native-oxide defined ring contact for low threshold vertical-cavity lasers", *Appl. Phys. Lett.*, vol. 65, pp. 97–99, 1994.

[69] M.H. MacDougal, H. Zhao, P.D. Dapkus, M. Ziari, and W.H. Steier, "Wide-bandwidth distributed Bragg reflectors using oxide/GaAs multilayers", *Electron. Lett.*, vol. 30, pp. 1147–1148, 1994.

[70] S. Adachi, "Optical Properties of AlGaAs: Transparent and Interband-Transition Regions (Tables)," Chap. 5 in *Properties of Aluminium Gallium Arsenide*, S. Adachi (Ed.), pp. 125–140. London, UK: INSPEC, 1993.

[71] M. Grabherr, R. Jäger, R. Michalzik, B. Weigl, G. Reiner, and K.J. Ebeling, "Efficient single-mode oxide-confined GaAs VCSEL's emitting in the 850-nm wavelength regime", *IEEE Photon. Technol. Lett.*, vol. 9, pp. 1304–1306, 1997.

[72] B. Weigl, M. Grabherr, C. Jung, R. Jäger, G. Reiner, R. Michalzik, D. Sowada, and K.J. Ebeling, "High-performance oxide-confined GaAs VCSELs", *IEEE J. Select. Topics Quantum Electron.*, vol. 3, pp. 409–415, 1997.

[73] W.P. Dumke, "Interband transitions and maser action", *Phys. Rev.*, vol. 127, pp. 1559–1563, 1962.

[74] W. Nakwaski and M. Osiński, "Thermal resistance of top-surface-emitting vertical-cavity semiconductor lasers and monolithic two-dimensional arrays", *Electron. Lett.*, vol. 28, pp. 572–574, 1992. Corrected in *Electron. Lett.*, vol. 28, p. 1283, 1992.

[75] S. Adachi, "Lattice thermal conductivity of group-IV and III-V semiconductor alloys", *J. Appl. Phys.*, vol. 102, pp. 063502-1–7, 2007.

[76] S. Adachi, "Thermal Properties", Chap. 3 in *Properties of Aluminium Gallium Arsenide*, S. Adachi (Ed.), pp. 37–49. London, UK: INSPEC, 1993.

[77] K.Y. Lau, "Dynamics of Quantum Well Lasers", Chap. 5 in *Quantum Well Lasers*, P.S. Zory, Jr. (Ed.). San Diego, USA: Academic Press, 1993.

[78] W. Rideout, W.F. Sharfin, E.S. Koteles, M.O. Vassell, and B. Elman, "Well-barrier hole burning in quantum well lasers", *IEEE Photon. Technol. Lett.*, vol. 3, pp. 784–786, 1991.

[79] L.A. Coldren and S.W. Corzine, *Diode Lasers and Photonic Integrated Circuits*. New York, USA: J. Wiley & Sons, 1995.

[80] A. Al-Samaneh, M. Bou Sanayeh, S. Renz, D. Wahl, and R. Michalzik, "Polarization control and dynamic properties of VCSELs for MEMS atomic clock applications", *IEEE Photon. Technol. Lett.*, vol. 23, pp. 1049–1051, 2011.

[81] K.L. Lear and A.N. Al-Omari, "Progress and issues for high speed vertical cavity surface emitting lasers", in *Vertical-Cavity Surface-Emitting Lasers XI*, K.D. Choquette and J.K. Guenter (Eds.), Proc. SPIE 6484, pp. 64840J-1–12, 2007.

[82] K.L. Lear, V.M. Hietala, H.Q. Hou, J. Banas, B.E. Hammons, J. Zolper, S.P. Kilcoyne, "Small and large signal modulation of 850 nm oxide-confined vertical-cavity surface-emitting lasers", in Proc. *Conference on Lasers and Electro-Optics, 1997, CLEO '97*, pp. 193–194. Baltimore, USA, May 1997.

[83] K.L. Lear, V.M. Hietala, H.Q. Hou, M. Ochiai, J.J. Banas, B.E. Hammons, J.C. Zolper, and S.P. Kilcoyne, "Small and large signal modulation of 850 nm oxide-confined vertical cavity surface emitting lasers", *OSA Trends in Optics and Photonics*, vol. 15, *Advances in Vertical Cavity Surface Emitting Lasers*, C.J. Chang-Hasnain (Ed.), pp. 69–74, 1997.

[84] P.V. Mena, J.J. Morikuni, S.-M. Kang, A.V. Harton, and K.W. Wyatt, "A comprehensive circuit-level model of vertical-cavity surface-emitting lasers", *J. Lightwave Technol.*, vol. 17, pp. 2612–2632, 1999.

[85] J.W. Scott, R.S. Geels, S.W. Corzine, and L.A. Coldren, "Modeling temperature effects and spatial hole burning to optimize vertical-cavity surface-emitting laser performance", *IEEE J. Quantum Electron.*, vol. 29, pp. 1295–1308, 1993.

[86] O. Kjebon, R. Schatz, S. Lourdudoss, S. Nilsson, and B. Stalnacke, "Modulation response measurements and evaluation of MQW InGaAsP lasers of various designs", in *High-Speed Semiconductor Laser Sources*, P.A. Morton and D.L. Crawford (Eds.), Proc. SPIE 2684, pp. 138–152, 1996.

[87] A.N. Al-Omari, I.K. Al-Kofahi, and K.L. Lear, "Fabrication, performance and parasitic parameter extraction of 850 nm high-speed vertical-cavity lasers", *Semicond. Sci. Technol.*, vol. 24, pp. 095024-1–8, 2009.

[88] C.H. Henry, "Theory of the linewidth of semiconductor lasers", *IEEE J. Quantum Electron.*, vol. 18, pp. 259–264, 1982.

[89] K.J. Ebeling, *Integrated Optoelectronics*. Berlin, Germany: Springer, 1993.

[90] R. Wynands and A. Nagel, "Inversion of frequency-modulation spectroscopy line shapes", *J. Opt. Soc. Am. B*, vol. 16, pp. 1617–1622, 1999.

[91] C. Harder, K. Vahala, and A. Yariv, "Measurement of the linewidth enhancement factor α of semiconductor lasers", *Appl. Phys. Lett.*, vol. 42, pp. 328–330, 1983.

[92] C.M. Long and K.D. Choquette, "Optical characterization of a vertical cavity surface emitting laser for a coherent population trapping frequency reference", *J. Appl. Phys.*, vol. 103, pp. 033101-1–5, 2008.

[93] K. Petermann, *Laser Diode Modulation and Noise*. Dordrecht, The Netherlands: Kluwer Academic Publishers, 1991.

[94] M.C. Tatham, I.F. Lealman, C.P. Seltzer, L.D. Westbrook, and D.M. Cooper, "Resonance frequency, damping, and differential gain in 1.5 µm multiple quantum-well lasers", *IEEE J. Quantum Electron.*, vol. 28, pp. 408–414, 1992.

[95] W. Schmid, C. Jung, B. Weigl, G. Reiner, R. Michalzik, and K.J. Ebeling, "Delayed self-heterodyne linewidth measurement of VCSEL's", *IEEE Photon. Technol. Lett.*, vol. 8, pp. 1288–1290, 1996.

[96] D.K. Serkland, G.A. Keeler, K.M. Geib, and G.M. Peake, "Narrow linewidth VCSELs for high-resolution spectroscopy", in *Vertical-Cavity Surface-Emitting Lasers XIII*, K.D. Choquette and C. Lei (Eds.), Proc. SPIE 7229, pp. 722907-1–8, 2009.

[97] S. Barland, P. Spinicelli, G. Giacomelli, and F. Marin, "Measurement of the working parameters of an air-post vertical-cavity surface-emitting laser", *IEEE J. Quantum Electron.*, vol. 41, pp. 1235–1243, 2005.

[98] D. Kuksenkov, S. Feld, C. Wilmsen, H. Temkin, S. Swirhun, and R. Leibenguth, "Linewidth and α-factor in AlGaAs/GaAs vertical cavity surface emitting lasers", *Appl. Phys. Lett.*, vol. 66, pp. 277–279, 1995.

[99] G. Di Domenico, S. Schilt, and P. Thomann, "Simple approach to the relation between laser frequency noise and laser line shape", *Appl. Opt.*, vol. 49, pp. 4801–4807, 2010.

[100] M.P. van Exter, A.K. Jansen van Doorn, and J.P. Woerdman, "Electro-optic effect and birefringence in semiconductor vertical-cavity lasers", *Phys. Rev. A*, vol. 56, pp. 845–853, 1997.

[101] M.P. van Exter, R.F.M. Hendriks, and J.P. Woerdman, "Physical insight into the polarization dynamics of semiconductor vertical-cavity lasers", *Phys. Rev. A*, vol. 57, pp. 2080–2090, 1998.

[102] K.D. Choquette, K.M. Geib, C.I.H. Ashby, R.D. Twesten, O. Blum, H.Q. Hou, D.M. Follstaedt, B.E. Hammons, D. Mathes, and R. Hull, "Advances in selective wet oxidation of AlGaAs alloys", *IEEE J. Select. Topics Quantum Electron.*, vol. 3, pp. 916–926, 1997.

[103] M. Peeters, K. Panajotov, G. Verschaffelt, B. Nagler, J. Albert, H. Thienpont, I. Veretennicoff, and J. Danckaert, "Polarisation behavior of vertical-cavity surface-emitting lasers under the influence of in-plane anisotropic strain", in *Vertical-Cavity Surface-Emitting Lasers VI*, C. Lei and S.P. Kilcoyne (Eds.), Proc. SPIE 4649, pp. 281–291, 2002.

[104] A. Yariv and P. Yeh, *Optical Waves in Crystals*. New York, USA: John Wiley & Sons, 1995.

[105] K. Panajotov, B. Ryvkin, J. Danckaert, M. Peeters, H. Thienpont, and I. Veretennicoff, "Polarization switching in VCSEL's due to thermal lensing", *IEEE Photon. Technol. Lett.*, vol. 10, pp. 6–8, 1998.

Bibliography

[106] A. Valle, L. Pesquera, and K.A. Shore, "Polarization behavior of birefringent multitransverse mode vertical-cavity surface-emitting lasers", *IEEE Photon. Technol. Lett.*, vol. 9, pp. 557–559, 1997.

[107] T. Ackemann and M. Sondermann, "Characteristics of polarization switching from the low to the high frequency mode in vertical-cavity surface-emitting lasers", *Appl. Phys. Lett.*, vol. 78, pp. 3574–3576, 2001.

[108] T. Mukaihara, N. Ohnoki, Y. Hayashi, N. Hatori, F. Koyama, and K. Iga, "Excess intensity noise originated from polarization fluctuation in vertical-cavity surface-emitting lasers", *IEEE Photon. Technol. Lett.*, vol. 7, pp. 1113–1115, 1995.

[109] G. Totsching, M. Lackner, R. Shau, M. Ortsiefer, J. Rosskopf, M.C. Amann, and F. Winter, "High-speed vertical-cavity surface-emitting laser (VCSEL) absorption spectroscopy of ammonia (NH_3) near 1.54 µm", *Appl. Phys. B*, vol. 76, pp. 603–608, 2003.

[110] B. Scherer, J. Wöllenstein, M. Weidemüller, W. Salzmann, J.M. Ostermann, F. Rinaldi, and R. Michalzik, "Oxygen measurements at high pressures using a low cost, polarisation stabilized, widely tunable vertical-cavity surface-emitting laser", in *Smart Sensors, Actuators, and MEMS III*, T. Becker, C. Cané, and N.S. Barker (Eds.), Proc. SPIE 6589, pp. 658908-1–10, 2007.

[111] A. Pruijmboom, M. Schemmann, J. Hellmig, J. Schutte, H. Moench, and J. Pankert, "VCSEL-based miniature laser-Doppler interferometer", in *Vertical-Cavity Surface-Emitting Lasers XII*, C. Lei and J.K. Guenter (Eds.), Proc. SPIE 6908, pp. 69080I-1–7, 2008.

[112] A. Pruijmboom, S. Booij, M. Schemmann, K. Werner, P. Hoeven, H. van Limpt, S. Intemann, R. Jordan, T. Fritzsch, H. Oppermann, and M. Barge, "VCSEL-based miniature laser-self-mixing interferometer with integrated optical and electronic components", in *Photonics Packaging, Integration, and Interconnects IX*, A.L. Glebov and R.T. Chen (Eds.), Proc. SPIE 7221, pp. 72210S-1–12, 2009.

[113] M. Grabherr, R. King, R. Jäger, D. Wiedenmann, P. Gerlach, D. Duckeck, and C. Wimmer, "Volume production of polarization-controlled single-mode VCSELs", in *Vertical-Cavity Surface-Emitting Lasers XII*, C. Lei and J.K. Guenter (Eds.), Proc. SPIE 6908, pp. 690803-1–9, 2008.

[114] M. Grabherr, H. Moench, and A. Pruijmboom, "VCSELs for Optical Mice and Sensing", Chap. 17 in *VCSELs — Fundamentals, Technology and Applications of Vertical-Cavity Surface-Emitting Lasers*, R. Michalzik (Ed.), Springer Series in Optical Sciences, vol. 166, pp. 521–538. Berlin, Germany: Springer-Verlag, 2013.

[115] J.A. Tatum, "VCSEL proliferation", in *Vertical-Cavity Surface-Emitting Lasers XI*, K.D. Choquette and J.K. Guenter (Eds.), Proc. SPIE 6484, pp. 648403-1–7, 2007.

[116] K.P. Jackson and C.L. Schow, "VCSEL-Based Transceivers for Data Communications", Chap. 14 in *VCSELs — Fundamentals, Technology and Applications of Vertical-Cavity*

Surface-Emitting Lasers, R. Michalzik (Ed.), Springer Series in Optical Sciences, vol. 166, pp. 431–448. Berlin, Germany: Springer-Verlag, 2013.

[117] S.T. Fard, W. Hofmann, P.T. Fard, G. Böhm, M. Ortsiefer, E. Kwok, M.C. Amann, and L. Chrostowski, "Optical absorption glucose measurements using 2.3 µm vertical cavity semiconductor laser", *IEEE Photon. Technol. Lett.*, vol. 20, pp. 930–932, 2008.

[118] M.E. Webber, *Diode Laser Measurements of NH_3 and CO_2 for combustion and bioreactor applications*. Ph.D. thesis, Stanford University, Stanford, USA, 2001. Published online: http://hanson.stanford.edu/dissertations/Webber_2001.pdf, last visited October 2013.

[119] M. Ortsiefer, W. Hofmann, J. Rosskopf and M.C. Amann, "Long-Wavelength VCSELs with Buried Tunnel Junction", Chap. 10 in *VCSELs — Fundamentals, Technology and Applications of Vertical-Cavity Surface-Emitting Lasers*, R. Michalzik (Ed.), Springer Series in Optical Sciences, vol. 166, pp. 321–351. Berlin, Germany: Springer-Verlag, 2013.

[120] J.M. Ostermann, F. Rinaldi, P. Debernardi, and R. Michalzik, "VCSELs with enhanced single-mode power and stabilized polarization for oxygen sensing", *IEEE Photon. Technol. Lett.*, vol. 17, pp. 2256–2258, 2005.

[121] A. Al-Samaneh, S. Renz, A. Strodl, W. Schwarz, D. Wahl, and R. Michalzik, "Polarization-stable single-mode VCSELs for Cs-based MEMS atomic clock applications", in *Semiconductor Lasers and Laser Dynamics IV*, K. Panajotov, M. Sciamanna, A.A. Valle, and R. Michalzik (Eds.), Proc. SPIE 7720, pp. 772006-1–14, 2010.

[122] H. Otoma, A. Murakami, Y. Kuwata, N. Ueki, N. Mukoyama, T. Kondo, A. Sakamoto, S. Omori, H. Nakayama, and T. Nakamura, "Single-mode oxide-confined VCSEL for printers and sensors", in Proc. *Electronics System Integration Technology Conference*, pp. 80–85. Dresden, Germany, Sep. 2006.

[123] N. Mukoyama, H. Otoma, J. Sakurai, N. Ueki, and H. Nakayama, "VCSEL array based light exposure system for laser printing", in *Vertical-Cavity Surface-Emitting Lasers XII*, C. Lei and J.K. Guenter (Eds.), Proc. SPIE 6908, pp. 69080H-1–11, 2008.

[124] N. Ueki and N. Mukoyama, "VCSEL-Based Laser Printing System", Chap. 18 in *VCSELs — Fundamentals, Technology and Applications of Vertical-Cavity Surface-Emitting Lasers*, R. Michalzik (Ed.), Springer Series in Optical Sciences, vol. 166, pp. 539–548. Berlin, Germany: Springer-Verlag, 2013.

[125] B. Shao, S. Zlatanovic, and S.C. Esener, "Microscope-integrated micromanipulator based on multiple VCSEL traps", in *Optical Trapping and Optical Micromanipulation*, K. Dholakia and G.C. Spalding (Eds.), Proc. SPIE 5514, pp. 62–72, 2004.

[126] A. Bergmann, N.I. Khan, J.A. Martos Calahorro, D. Wahl, and R. Michalzik, "Hybrid integration approach of VCSELs for miniaturized optical deflection of microparticles", in *Semiconductor Lasers and Laser Dynamics V*, K.P. Panajotov, M. Sciamanna, A.A. Valle, and R. Michalzik (Eds.), Proc. SPIE 8432, pp. 843204-1–12, 2012.

Bibliography

[127] S. Adachi, "GaAs, AlAs, and $Al_xGa_{1-x}As$: Material parameters for use in research and device applications", *J. Appl. Phys.*, vol. 58, pp. R1–R29, 1985.

[128] H.C. Casey, Jr. and M.B. Panish, *Heterostructure Lasers, Part A: Fundamental Principles*. Orlando, USA: Academic Press, 1978.

[129] Y.P. Varshni, "Temperature dependence of the energy gap in semiconductors", *Physica*, vol. 34, pp. 149–154, 1967.

[130] S. Logothetidis, M. Cardona, and M. Garriga, "Temperature dependence of the dielectric function and the interband critical-point parameters of $Al_xGa_{1-x}As$", *Phys. Rev. B*, vol. 43, pp. 11950–11965, 1991.

[131] S. Paul, J.B. Roy, and P.K. Basu, "Empirical expressions for the alloy composition and temperature dependence of the band gap and intrinsic carrier density in $Ga_xIn_{1-x}As$", *J. Appl. Phys.*, vol. 69, pp. 827–829, 1991.

[132] P. Lautenschlager, M. Garriga, S. Logothetidis, and M. Cardona, "Interband critical points of GaAs and their temperature dependence", *Phys. Rev. B*, vol. 35, pp. 9174–9189, 1987.

[133] Z.M. Fang, K.Y. Ma, D.H. Jaw, R.M. Cohen, and G.B. Stringfellow, "Photoluminescence of InSb, InAs, and InAsSb grown by organometallic vapor phase epitaxy", *J. Appl. Phys.*, vol. 67, pp. 7034–7039, 1990.

[134] J. Singh, *Physics of Semiconductors and Their Heterostructures*. New York, USA: McGraw-Hill, 1993.

[135] P.S. Zory, *Quantum Well Lasers*. San Diego, USA: Academic Press, 1993.

[136] P. Harrison, *Quantum Wells, Wires and Dots—Theoretical and Computational Physics of Semiconductor Nanostructures*. Chichester, England: John Wiley & Sons, 2005.

[137] I. Vurgaftman, J.R. Meyer, and L.R. Ram-Mohan, "Band parameters for III–V compound semiconductors and their alloys", *J. Appl. Phys.*, vol. 89, pp. 5815–5875, 2001.

[138] D. Setz, *Entwicklung und Herstellung von Vertikallaserdioden zur Nutzung in Atomuhren*, Diploma thesis, Ulm University, Inst. of Optoelectronics, Ulm, Germany, Sept. 2008.

[139] D.J. Bendaniel and C.B. Duke, "Space-charge effects on electron tunneling", *Phys. Rev.*, vol. 152, pp. 683–692, 1966.

[140] J.W. Conley, C.B. Duke, G.D. Mahan, and J.J. Tiemann, "Electron tunneling in metal-semiconductor barriers", *Phys. Rev.*, vol. 150, pp. 466–469, 1966.

[141] S. Adachi, "Material parameters of $In_{1-x}Ga_xAs_yP_{1-y}$ and related binaries", *J. Appl. Phys.*, vol. 53, pp. 8775–8792, 1982.

[142] T.C. Chong and C.G. Fonstad, "Theoretical gain of strained-layer semiconductor lasers in the large strain regime", *IEEE J. Quantum Electron.*, vol. 25, pp. 171–178, 1989.

[143] H. Haug and S.W. Koch, "Semiconductor laser theory with many-body effects", *Phys. Rev. A*, vol. 39, pp. 1887–1898, 1989.

Bibliography

[144] R. Michalzik, *Modellierung und Design von Laserdioden mit Vertikalresonator*, Ph.D. thesis, Ulm University, Ulm, Germany, 1996. In: Fortschritt-Berichte, VDI-Reihe 9, Nr. 257. Düsseldorf, Germany: VDI Verlag, 1997.

[145] H.J. Unold, S.W.Z. Mahmoud, R. Jäger, M. Grabherr, R. Michalzik, and K.J. Ebeling, "Large-area single-mode VCSELs and the self-aligned surface relief", *IEEE J. Select. Topics Quantum Electron.*, vol. 2, pp. 386–392, 2001.

[146] A. Kroner, F. Rinaldi, J.M. Ostermann, and R. Michalzik, "High-performance single fundamental mode AlGaAs VCSELs with mode-selective mirror reflectivities", *Optics Communications*, vol. 270, pp. 332–335, 2007.

[147] Å. Haglund, J.S. Gustavsson, J.A. Vukušić, P. Modh, and A. Larsson, "Single fundamental-mode output power exceeding 6 mW from VCSELs with a shallow surface relief", *IEEE Photon. Technol. Lett.*, vol. 16, pp. 368–370, 2004.

[148] A. Larsson and J.S. Gustavsson, "Single-Mode VCSELs", Chap. 4 in *VCSELs — Fundamentals, Technology and Applications of Vertical-Cavity Surface-Emitting Lasers*, R. Michalzik (Ed.), Springer Series in Optical Sciences, vol. 166, pp. 119–144. Berlin, Germany: Springer-Verlag, 2013.

[149] E.W. Young, K.D. Choquette, S.L. Chuang, K.M. Geib, A.J. Fischer, and A.A. Allerman, "Single-transverse-mode vertical-cavity lasers under continuous and pulsed operation", *IEEE Photon. Technol. Lett.*, vol. 13, pp. 927–929, 2001.

[150] H.J. Unold, M.C. Riedl, S.W.Z. Mahmoud, R. Jäger, and K.J. Ebeling, "Long monolithic cavity VCSELs for high singlemode output power", *Electron. Lett.*, vol. 37, pp. 178–179, 2001.

[151] J.-W. Shi, C.-C. Chen, Y.-S. Wu, S.-H. Guol, C. Kuo, and Y.-J. Yang, "High-power and high-speed Zn-diffusion single fundamental-mode vertical-cavity surface-emitting lasers at 850-nm wavelength", *IEEE Photon. Technol. Lett.*, vol. 20, pp. 1121–1123, 2008.

[152] K.D. Choquette and R.E. Leibenguth, "Control of vertical-cavity laser polarization with anisotropic transverse cavity geometries", *IEEE Photon. Technol. Lett.*, vol. 8, pp. 40–42, 1994.

[153] T. Yoshikawa, H. Kosaka, K. Kurihara, M. Kajita, Y. Sugimoto, and K. Kasahara, "Complete polarization control of 8 × 8 vertical-cavity surface-emitting laser matrix arrays", *Appl. Phys. Lett.*, vol. 66, 908–910, 1994.

[154] H.Y. Chu, B.-S. Yoo, M.S. Park, and H.-H. Park, "Polarization characteristics of index-guided surface-emitting lasers with tilted pillar structure", *IEEE Photon. Technol. Lett.*, vol. 9, pp. 1066–1068, 1997.

[155] T. Yoshikawa, T. Kawakami, H. Saito, H. Kosaka, M. Kajita, K. Kurihara, Y. Sugimoto, and K. Kasahara, "Polarization-controlled single-mode VCSEL", *IEEE J. Quantum Electron.*, vol. 34, pp. 1009–1015, 1998.

Bibliography

[156] M. Takahashi, N. Egami, T. Mukaihara, F. Koyama, and K. Iga, "Lasing characteristics of GaAs(311)A substrate based InGaAs-GaAs vertical-cavity surface-emitting lasers", *IEEE J. Select. Topics Quantum Electron.*, vol. 3, pp. 372–378, 1997.

[157] M. Takahashi, P. Vaccaro, N. Egami, A. Mizutani, A. Matsutani, F. Koyama, and K. Iga, "Oxide-confinement vertical-cavity surface-emitting lasers grown on GaAs (311)A substrates with dynamically stable polarisation", *Electron. Lett.*, vol. 34, pp. 276–278, 1998.

[158] A. Mizutani, N. Hatori, N. Nishiyama, F. Koyama, and K. Iga, "A low-threshold polarization-controlled vertical-cavity surface-emitting laser grown on GaAs (311)B substrate", *IEEE Photon. Technol. Lett.*, vol. 10, pp. 633–635, 1998.

[159] A. Mizutani, N. Hatori, N. Nishiyama, F. Koyama, and K. Iga, "MOCVD grown InGaAs/GaAs vertical cavity surface emitting laser on GaAs (311)B substrate", *Electron. Lett.*, vol. 33, pp. 1877–1878, 1997.

[160] K. Tateno, Y. Ohiso, C. Amano, A. Wakatsuki, and T. Kurokawa, "Growth of verticalcavity surface-emitting laser structures on GaAs (311)B substrates by metalorganic chemical vapor deposition", *Appl. Phys. Lett.*, vol. 70, pp. 3395–3397, 1997.

[161] N. Nishiyama, A. Mizutani, N. Hatori, M. Arai, F. Koyama, and K. Iga, "Lasing characteristics of InGaAs-GaAs polarization controlled vertical-cavity surface-emitting laser grown on GaAs (311)B substrate", *IEEE J. Select. Topics Quantum Electron.*, vol. 5, pp. 530–536, 1999.

[162] C.I. Wilkinson, J. Woodhead, J.E.F. Frost, J.S. Roberts, R. Wilson, and M.F. Lewis, "Electrical polarization control of vertical-cavity surface-emitting lasers using polarized feedback and a liquid crystal", *IEEE Photon. Technol. Lett.*, vol. 11, pp. 155–157, 1999.

[163] M. Arizaleta Arteaga, M. López-Amo, H. Thienpont, and K. Panajotov, "Tailoring light polarization in vertical cavity surface emitting lasers by isotropic optical feedback from an extremely short external cavity", *Appl. Phys. Lett.*, vol. 89, pp. 091102-1–3, 2006.

[164] M. Arizaleta Arteaga, O. Parriaux, M. López-Amo, H. Thienpont, and K. Panajotov, "Polarized optical feedback from an extremely short external cavity for controlling and stabilizing the polarization of vertical cavity surface emitting lasers", *Appl. Phys. Lett.*, vol. 90, pp. 121104-1–3, 2007.

[165] T. Mukaihara, N. Ohnoki, Y. Hayashi, N. Hatori, F. Koyama, and K. Iga, "Polarization control of vertical-cavity surface-emitting lasers using a birefringent metal/dielectric polarizer loaded on top distributed Bragg reflector", *IEEE J. Select. Topics Quantum Electron.*, vol. 1, pp. 667–673, 1995.

[166] J.-H. Ser, Y.-G. Ju, J.-H. Shin, and Y.-H. Lee, "Polarization stabilization of vertical-cavity top-surface-emitting lasers by inscription of fine metal-interlaced gratings", *Appl. Phys. Lett.*, vol. 66, pp. 2769–2771, 1995.

[167] S. Goeman, S. Boons, B. Dhoedt, K. Vandeputte, K. Caekebeke, P. Van Daele, and R. Baets, "First demonstration of highly reflective and highly polarization selective diffraction gratings (GIRO-gratings) for long-wavelength VCSELs", *IEEE Photon. Technol. Lett.*, vol. 10, pp. 1205–1207, 1998.

[168] C.-A. Berseth, B. Dwir, I. Utke, H. Pier, A. Rudra, V.P. Iakovlev, and E. Kapon, "Vertical cavity surface emitting lasers incorporating structured mirrors patterned by electron-beam lithography", *J. Vac. Sci. Technol. B*, vol. 17, pp. 3222–3225, 1999.

[169] H.J. Unold, M.C. Riedl, R. Michalzik, and K.J. Ebeling, "Polarisation control in VCSELs by elliptic surface etching", *Electron. Lett.*, vol. 38, pp. 77–78, 2002.

[170] P. Debernardi, H.J. Unold, J. Maehnss, R. Michalzik, G.P. Bava, and K.J. Ebeling, "Single-mode, single-polarization VCSELs via elliptical surface etching: experiments and theory", *IEEE J. Select. Topics Quantum Electron.*, vol. 9, pp. 1394–1404, 2003.

[171] K.-H. Lee, J.-H. Baek, I.-K. Hwang, Y.-H. Lee, G.-H. Lee, J.-H. Ser, H.-D. Kim, and H.-E. Shin, "Square-lattice photonic-crystal vertical-cavity surface-emitting lasers", *Opt. Express*, vol. 12, pp. 4136–4143, 2004.

[172] P. Debernardi and G.P. Bava, "Coupled mode theory: a powerful tool for analyzing complex VCSELs and designing advanced device features", *IEEE J. Select. Topics Quantum Electron.*, vol. 9, pp. 905–917, 2003.

[173] M.D. Austin, H. Ge, W. Wu, M. Li, Z. Yu, D. Wasserman, S.A. Lyon, and S.Y. Chou, "Fabrication of 5 nm linewidth and 14 nm pitch features by nanoimprint lithography", *Appl. Phys. Lett.*, vol. 84, pp. 5299–5301, 2004.

[174] R. Michalzik, J.M. Ostermann, and P. Debernardi, "Polarization-stable monolithic VCSELs" (invited), in *Vertical-Cavity Surface-Emitting Lasers XII*, C. Lei and J.K. Guenter (Eds.), Proc. SPIE 6908, pp. 69080A-1–16, 2008.

[175] J.M. Ostermann, P. Debernardi, C. Jalics, A. Kroner, M.C. Riedl, and R. Michalzik, "Monolithic polarization control of multimode VCSELs by a dielectric surface grating", in *Vertical-Cavity Surface-Emitting Lasers VIII*, C. Lei and S.P. Kilcoyne (Eds.), Proc. SPIE 5364, pp. 201–212, 2004.

[176] J.M. Ostermann, P. Debernardi, C. Jalics, A. Kroner, M.C. Riedl, and R. Michalzik, "Surface gratings for polarization control of single- and multi-mode oxide-confined vertical-cavity surface-emitting lasers", *Optics Communications*, vol. 246, pp. 511–519, 2005.

[177] P. Debernardi, J.M. Ostermann, M. Feneberg, C. Jalics, and R. Michalzik, "Reliable polarization control of VCSELs through monolithically integrated surface gratings: a comparative theoretical and experimental study", *IEEE J. Select. Topics Quantum Electron.*, vol. 11, pp. 107–116, 2005.

[178] P. Debernardi, "Three-Dimensional Modeling of VCSELs", Chap. 3 in *VCSELs — Fundamentals, Technology and Applications of Vertical-Cavity Surface-Emitting Lasers*, R.

Michalzik (Ed.), Springer Series in Optical Sciences, vol. 166, pp. 77–117. Berlin, Germany: Springer-Verlag, 2013.

[179] A. Al-Samaneh, M.T. Haidar, D. Wahl, and R. Michalzik, "Polarization-stable single-mode VCSELs for Cs-based miniature atomic clocks", in Online Digest *Conf. on Lasers and Electro-Optics Europe, CLEO/Europe 2011*, paper CB.P.23, one page. Munich, Germany, May 2011.

[180] P. Debernardi, J.M. Ostermann, M. Sondermann, T. Ackemann, G.P. Bava, and R. Michalzik, "Theoretical-experimental study of the vectorial modal properties of polarization-stable multimode grating VCSELs", *IEEE J. Select. Topics Quantum Electron.*, vol. 13, pp. 1340–1348, 2007.

[181] J.M. Ostermann, P. Debernardi, and R. Michalzik, "Surface grating VCSELs with dynamically stable light output polarization", *IEEE Photon. Technol. Lett.*, vol. 17, pp. 2505–2507, 2005.

[182] J.M. Ostermann, P. Debernardi, A. Kroner, and R. Michalzik, "Polarization-controlled surface grating VCSELs under externally induced anisotropic strain", *IEEE Photon. Technol. Lett.*, vol. 19, pp. 1301–1303, 2007.

[183] J.M. Ostermann, P. Debernardi, and R. Michalzik, "Polarization-controlled surface grating VCSELs under unpolarized and polarized optical feedback", *IEEE Photon. Technol. Lett.*, vol. 19, pp. 1359–1361, 2007.

[184] D.R. Pendse, A.K. Chin, F.P. Dabkowski, and E.M. Clausen, "Reliability comparison of GaAlAs/GaAs and aluminum-free high-power laser diodes", in *Semiconductor Lasers III*, S. Forouhar, Q. Wang, and L.J. Davis (Eds.), Proc. SPIE 3547, pp. 79–85, 1998.

[185] J.M. Ostermann, P. Debernardi, C. Jalics, and R. Michalzik, "Shallow surface gratings for high-power VCSELs with one preferred polarization for all modes", *IEEE Photon. Technol. Lett.*, vol. 17, pp. 1593–1595, 2005.

[186] J.M. Ostermann, P. Debernardi, and R. Michalzik, "Optimized integrated surface grating design for polarization-stable VCSELs", *IEEE J. Quantum Electron.*, vol. 42, pp. 690–698, 2006.

[187] J.M. Ostermann, P. Debernardi, C. Jalics, and R. Michalzik, "Polarization-stable oxide-confined VCSELs with enhanced single-mode output power via monolithically integrated inverted grating reliefs", *IEEE J. Select. Topics Quantum Electron.*, vol. 11, pp. 982–989, 2005.

[188] A. Al-Samaneh, M. Bou Sanayeh, M.J. Miah, W. Schwarz, D. Wahl, A. Kern, and R. Michalzik, "Polarization-stable vertical-cavity surface-emitting lasers with inverted grating relief for use in microscale atomic clocks", *Appl. Phys. Lett.*, vol. 101, pp. 171104-1–4, 2012.

[189] M. Sugimoto, H. Kosaka, K. Kurihara, I. Ogura, T. Numai, and K. Kasahara, "Very low threshold current density in vertical-cavity surface-emitting laser diodes with periodically doped distributed Bragg reflectors", *Electron. Lett.*, vol. 28, pp. 385–387, 1992.

[190] R.S. Geels, S.W. Corzine, J.W. Scott, D.B. Young, and L.A. Coldren, "Low threshold planarized vertical-cavity surface-emitting lasers", *IEEE Photon. Technol. Lett.*, vol. 2, pp. 234–236, 1990.

[191] D. Wiedenmann, M. Grabherr, R. Jäger, and R. King, "High volume production of single-mode VCSELs", in *Vertical-Cavity Surface-Emitting Lasers X*, C. Lei and K.D. Choquette (Eds.), Proc. SPIE 6132, pp. 613202-1–12, 2006.

[192] S. Guha, F. Agahi, B. Pezeshki, J.A. Kash, D.W. Kisker, and N.A. Bojarczuk, "Microstructure of AlGaAs-oxide heterolayers formed by wet oxidation", *Appl. Phys. Lett.*, vol. 68, pp. 906–909, 1996.

[193] K.D. Choquette, K.M. Geib, H.C. Chui, B.E. Hammons, H.Q. Hou, T.J. Drummond, and R. Hull, "Selective oxidation of buried AlGaAs versus AlAs layers", *Appl. Phys. Lett.*, vol. 69, pp. 1385–1387, 1996.

[194] M.A. McCord and M.J. Rooks, "Electron beam lithography", Chap. 2 in *Handbook of Microlithography, Micromachining, and Microfabrication. Volume 1: Microlithograph*, P. Rai-Choudhury (Ed.), pp. 139–250. Washington DC, USA: SPIE Press Monograph, 1997.

[195] M. Parikh and D.F. Kyser, "Energy deposition functions in electron resist films on substrates", *J. Appl. Phys.*, vol. 50, pp. 1104–1111, 1979.

[196] E. Kratschmer, "Verification of a proximity effect correction program in electron-beam lithography", *J. Vac. Sci. Technol.*, vol. 19, pp. 1264–1268, 1981.

[197] S.A. Rishton and D.P. Kern, "Point exposure distribution measurements for proximity correction in electron beam lithography on a sub-100 nm scale", *J. Vac. Sci. Technol. B*, vol. 5, pp. 135–141, 1987.

[198] G. Owen and P. Rissman, "Proximity effect correction for electron beam lithography by equalization of background dose", *J. Appl. Phys.*, vol. 54, pp. 3573–3581, 1983.

[199] G.C. DeSalvo, W.F. Tseng, and J. Comas, "Etch rates and selectivities of citric acid hydrogen peroxide on GaAs, $Al_{0.3}Ga_{0.7}As$, $In_{0.2}Ga_{0.8}As$, $In_{0.53}Ga_{0.47}As$, $In_{0.52}Al_{0.48}As$, and InP", *J. Electrochem. Soc.*, vol. 139, pp. 831–835, 1992.

[200] J.M. Ostermann, *Differactive Optics for Polarization Control of Vertical-Cavity Surface-Emitting Lasers*, Ph.D. thesis, Ulm University, Ulm, Germany, 2007. Göttingen, Germany: Cuvillier Verlag, 2007.

[201] R. Olshansky, P. Hill, V. Lanzisera, and W. Powazinik, "Frequency response of 1.3 µm InGaAsP high speed semiconductor lasers", *IEEE J. Quantum Electron.*, vol. QE-42, pp. 1410–1418, 1987.

[202] D. Wiedenmann, R. King, C. Jung, R. Jäger, R. Michalzik, P. Schnitzer, M. Kicherer, and K.J. Ebeling, "Design and analysis of single-mode oxidized VCSEL's for high-speed optical interconnects", *IEEE J. Select. Topics Quantum Electron.*, vol. 5, pp. 503–511, 1999.

Bibliography

[203] Y. Ou, J.S. Gustavsson, P. Westbergh, Å. Haglund, A. Larsson, and A. Joel, "Impedance characteristics and parasitic speed limitations of high-speed 850-nm VCSELs", *IEEE Photon. Technol. Lett.*, vol. 21, pp. 1840–1842, 2009.

[204] S.M. Sze and K.K. Ng, *Physics of Semiconductor Devices*, 3rd edition. New Jersey, USA: John Wiley and Sons, 2006.

[205] J. Cunningham, D. McElfresh, L. Lopez, D. Vacar, and A. Krishnamoorthy, "Scaling vertical-cavity surface-emitting laser reliability for petascale systems", *Appl. Opt.*, vol. 45, pp. 6342–6348, 2006.

[206] F. Gruet, A. Al-Samaneh, E. Kroemer, L. Bimboes, D. Miletic, C. Affolderbach, D. Wahl, R. Boudot, G. Mileti, and R. Michalzik, "Metrological characterization of custom-designed 894.6 nm VCSELs for miniature atomic clocks", *Opt. Express*, vol. 21, pp. 5781–5792, 2013.

[207] C. Affolderbach, A. Nagel, S. Knappe, C. Jung, D. Wiedenmann, and R. Wynands, "Nonlinear spectroscopy with a vertical-cavity surface-emitting laser", *Appl. Phys. B*, vol. 70, pp. 407–413, 2000.

[208] S. Kobayashi, Y. Yamamoto, M. Ito, and T. Kimura, "Direct frequency modulation in AlGaAs semiconductor lasers", *IEEE J. Quantum Electron.*, vol. 30, pp. 428–441, 1982.

[209] R.D. Cowan, *The Theory of Atomic Structure and Spectra*. Los Angeles, USA: University of California Press, 1981.

[210] D.A. Steck, "Cesium D Line Data", available online at http://steck.us/alkalidata (revision 2.1.4, Dec. 23, 2010).

[211] C.J. Foot, *Atomic Physics*. Oxford, UK: Oxford University Press, 2005.

[212] T. Udem, J. Reichert, R. Holzwarth, and T.W. Hänsch, "Absolute optical frequency measurement of the cesium D_1 line with a mode-locked laser", *Phys. Rev. Lett.*, vol. 82, pp. 3568–3571, 1999.

[213] V. Gerginov, A. Derevianko, and C.E. Tanner, "Observation of the nuclear magnetic octupole moment of ^{133}Cs", *Phys. Rev. Lett.*, vol. 91, pp. 072501-1-4, 2003.

[214] T. Udem, J. Reichert, T.W. Hänsch, and M. Kourogi, "Absolute optical frequency measurement of the cesium D_2 line", *Phys. Rev. A*, vol. 62, pp. 031801-1-4, 2000.

[215] P.A.M. Dirac, *The Principles of Quantum Mechanics*, 4th edition. Oxford, UK: Oxford University Press, 1958.

[216] S. Svanberg, *Atomic and Molecular Spectroscopy: Basic Aspects and Practical Applications*, 3rd edition. Berlin, Germany: Springer-Verlag, 2001.

[217] M. Kazu, *Charakterisierung von Laserdioden und Cs-Gaszellen für miniaturisierte Atomuhren*, Bachelor thesis, Ulm University, Inst. of Optoelectronics, Ulm, Germany, March 2013.

[218] P.J. Mohr, B.N. Taylor, and D.B. Newell, "CODATA recommended values of the fundamental physical constants: 2006", *Rev. Mod. Phys.*, vol. 80, pp. 633–730, 2008.

Bibliography

Curriculum Vitae

Personal Data
Ahmed Al-Samaneh
born on 24.08.1982 in Amman, Jordan

Employment

since 05/2014	R&D product design engineer, Avago Technologies, Regensburg, Germany
11/2013 – 04/2014	R&D photonics engineer, Innofactum, Osterhofen, Germany
07/2013 – 09/2013	Academic employee, Institute of Electron Devices and Circuits, Ulm University
10/2008 – 08/2012	Academic employee, Institute of Optoelectronics, Ulm University
06/2007 – 04/2008	Research assistant, Institute of Optoelectronics, Ulm University
10/2004 – 02/2006	Teaching assistant, Department of Electrical Engineering, Hashemite University, Jordan

Education

since 10/2008	PhD student in electrical engineering, Institute of Optoelectronics, Ulm University
05/2008	Master of Science in Communications Technology (M.Sc.), Ulm University
10/2007 – 05/2008	Master thesis, Institute of Optoelectronics, Ulm University
03/2006 – 05/2008	student in Communications Technology (master program), Ulm University
06/2004	Bachelor of Science in Electrical & Computer Engineering (B.Sc.), Hashemite University, Jordan
10/2000 – 06/2004	Student in Electrical & Computer Engineering, Hashemite University, Jordan
1988 – 2000	Al Ittihad Schools, Amman, Jordan

Scholarships and Awards

04/2010	Best Student Paper Award at the *Conf. on Semiconductor Lasers and Laser Dynamics IV* as part of *SPIE Photonics Europe* in Brussels, Belgium for the work on VCSELs for miniaturized atomic clocks
10/2007 – 04/2008	Six-months master thesis stipend awarded by Ulm University to the best students of the Communications Technology master program
05/2007	Best student award in the Communications Technology master program at Ulm University

Ulm, March 2014   Ahmed Al-Samaneh

www.ingramcontent.com/pod-product-compliance
Lightning Source LLC
Chambersburg PA
CBHW050208230526
45470CB00001B/289